Legal and Political Challenges of Governing the Environment and Climate Change

The environment has not always been protected by law. It was not until the middle of the twentieth century that 'the environment' came to be understood as an entity in need of special care, when the law–politics duo firmly fixed their focus on this issue.

In this book Wickham and Goodie tell the story of how law and politics first came upon the environment as an object in need of special attention. They outline the unlikely intersection of aesthetics and science that made 'the environment' into the matter of great concern it is today. The book describes the way private common-law strategies and public-law legislative strategies have approached the task of protecting the environment, and explore the greatest environmental challenge to have so far confronted environmental law and politics: the threat of global climate change. The book offers descriptions of many of the strategies being deployed to meet this challenge and presents some troubling assessments of them.

The book will be of great interest to students, teachers, and researchers of environmental law, socio-legal studies, environmental studies, and political theory.

Gary Wickham is Professor of Sociology at Murdoch University, Australia.

Jo-Ann Goodie is Senior Lecturer in Law at Murdoch University, Australia.

This excellent book explains environmental policy in general and climate change policy in particular as the product of what the authors call 'legal-political government'. The concept of legal-political government is an attempt to capture the nature of constitutionality and legality in liberal-democratic societies, and has the conditions of legitimate contestation over policy at its heart. The insights this concept brings to the successes and failure of environmental policy-making are striking and persuasive. This book should be read by all seeking some objective account of vexed environmental issues and particularly climate change.

Professor David Campbell, Law School, Lancaster University

Legal and Political Challenges of Governing the Environment and Climate Change

Ruling Nature

Gary Wickham and Jo-Ann Goodie

Routledge
Taylor & Francis Group

LONDON AND NEW YORK

First published 2013
by Routledge
2 Park Square, Milton Park, Abingdon, Oxfordshire OX14 4RN

and by Routledge
711 Third Avenue, New York, NY 10017

First issued in paperback 2015

Routledge is an imprint of the Taylor & Francis Group, an informa business

British Library Cataloguing in Publication Data
A catalogue record for this book is available from the British Library

Library of Congress Cataloging-in-Publication Data
Wickham, Gary, 1951-
Legal and political challenges of governing the environment and climate change : ruling nature / Gary
Wickham and Jo-Ann Goodie.
pages cm
Includes bibliographical references and index.
ISBN 978-0-415-67464-5 -- ISBN 978-0-203-79831-7 (e-book) 1. Environmental law. 2. Climatic
changes--Law and legislation. 3. Environmental responsibility. I. Goodie, Jo-Ann. II. Title.
K3585.5.W53 2013
344.04'6--dc23
2013025668

ISBN 13: 978-1-138-93754-3 (pbk)
ISBN 13: 978-0-415-67464-5 (hbk)

Typeset in 11/12 Garamond 3 by
Servis Filmsetting Ltd, Stockport, Cheshire

For MPHW, HNGDW, and JSW

Contents

Acknowledgements

As well as thanking Murdoch University for its research support, we wish to jointly thank the following people for their intellectually stimulating company and support over many years in many countries: Peter Baehr, Prudence Black, John Brigham, David Brown, Bernadette Cagan, David Campbell, Richard Collier, Paul du Gay, Barbara Evers, Farida Fozdar, David Garland, John Germov, Lee Godden, Anne Greenshields, Kirsten Harley, Christine Harrington, Patricia Harris, Barry Hindess, the late Paul Q. Hirst, Alan Hunt, Noel King, Martin Krygier, Hannah Lewi, Jeff Malpas, David McCallum, Stephen Muecke, Nick Osbaldiston, Pat O'Malley, George Pavlich, Sharyn Roach Anleu, Nikolas Rose, David Saunders, David Silverman, Wally Smith, Bill Taylor, Bryan Turner, Stephen Turner, Robert van Krieken, Ryan Walter, William Walters, Maximillian Wickham, and Lisa Young.

Gary Wickham wishes to express a special thanks to four people who have been remarkably patient and supportive mentors for over twenty-five years: Dick Bryan, Ian Hunter, Gavin Kendall, and Grahame Thompson.

We are grateful to the editors of *Law, Text, Culture* for their permission to reproduce material from: J. Goodie (2008) 'Toxic Tort and the Articulation of Environmental Risk' *Law Text and Culture* 12: 69–102.

1 Introduction

Towards a legal-political narrative

How can we know 'the environment'?

The environment is a chameleon. As an object of aesthetic contemplation it is landscape, horizon, and universe; it might be glimpsed through painting, poetry, novels, and sculpture but it is never totally exposed. As an object of morality it is both the source of natural good and a marker of humanity's capacity for degradation and exploitation. As an object of science it is an ever expanding and ever demanding frontier of knowledge – knowledge about what it was, what it is, and what it might be. As an object of economics it is the ultimate resource, that from which immense wealth can be created, but also that which can destroy wealth, precisely because it can destroy us. As an object of religion it is a gift from the heavens which confronts faith almost as often as it confirms it. As an object of law it is a tremendous challenge – the challenge of protecting it, as well as, among other things, the challenge of determining how it is to be protected and how its fruits are to be distributed. As an object of politics it is a site of constant contest.

In its capacity as a chameleon the environment is all these things simultaneously. This makes it anything but straightforward to write a book about how the environment and climate change are being governed by law and politics in a wide range of countries. We will fully explain this form of government in great detail in the next chapter. For now it is enough to know that we have in mind those countries which came to adopt the system of rule whereby law and politics are related in a very particular manner, a manner by which the two elements temper each other's attempts at superiority and in so doing end up producing a stronger whole than either part could ever have produced alone. The system emerged in the sixteenth and seventeenth centuries in England, France, Germany, and the Netherlands. It spread relatively quickly to other parts of Europe and to North America. It can now be said to include, at least, Australasia, much of South America, much of Asia, parts of Africa, and aspects of the United Nations and other multilateral forums.

In setting out to write the book we knew we could not simply excise the law–politics–environment and climate change intersection from its broader context, from its full personality, as it were. We knew we had to find a framework which would allow us to focus on law–politics–environment and climate change in such a way as to allow sufficient room for the aesthetic, moral, religious, scientific, and economic aspects of the environment and climate change. In other words, we had to make sure we could bring these other aspects into our law-and-politics story of the government of the environment and climate change.

We are confident we have found a way to meet the challenge. We propose that a particular slice of the history of natural law contains the seeds of those discourses that have made law and politics the dominant mechanisms for governing the environment and climate change in the wide range of countries set out above – dominant, but not so much so that they do not have to constantly deal with moral, religious, economic, aesthetic, and scientific elements as they go about the governing.

In saying this, we are not proposing a comprehensive history of natural law as a complex intersection of intersections – that is well beyond the scope of the book, and well beyond our competence. Instead, what we are offering – to be contained in a few pages of this Introduction – is a brief ideal-type account[1] of an important moment (albeit a long moment) in the history of natural law, a thumbnail sketch which can serve as the launching pad for the more substantive chapters of the book. Our sketch, we suggest, can help us explain, first, the way the environment has become an object of legal-political government, which is the main aim of our book,[2] and, second, to explain the way legal-political government is dealing with the threat of climate change, which is the subsidiary aim of our book.

An ideal-type account of a slice of the history of natural law as the basis of our account of the legal-political government of the environment and climate change

Our account of what we regard as the most crucial long moment in the history of natural law focuses more on the 'nature' component than the 'law' component. In short, our account has it that in Europe, from (at least) the thirteenth century through to the seventeenth, drawing on different ancient sources, two rival understandings of nature, contained in two rival understandings of human nature, developed in such a way as to underpin two rival versions of natural law, which in turn have gone on to underpin two rival approaches to

---

1 By ideal-type account we mean one consistent with Weber's 'self-proclaimed technique' of ideal types, which aims 'to intentionally emphasize certain facts in a "one sided way" as a method of analysis, but his point very often is also to conceptually define a domain or conceptual space in which even the most extreme actual cases have elements of the "opposite" conceptual category' (Turner 2002: 1).

2 In attempting to explain the emergence of the environment as an object of legal-political government perhaps we are attempting something like Ryan Walter's succinct explanation of the emergence of the economy as an object of what we would call legal-political government (Walter 2011).

government, one of which informs legal-political government. We will start with the two rival understandings of nature, which are contained in the two distinct understandings of human nature.

On the one hand is a vision of humans as a product of perfect nature. The earliest form of this vision was provided by Aristotle. For him, nature is perfect, so humans must have it in them to be perfect too, even though they more often than not think and act in extremely imperfect ways. The form of this explanation that has been most influential in the modern world is the Christianized form developed by the thirteenth-century theologian Thomas Aquinas, for whom God is the source of nature's perfection and hence the source of humans' potential for perfection. Aquinas's Christianized Aristotelianism, which became known as scholasticism, opened the door to later versions of nature's perfection, some of which, like that of Gottfried Wilhelm Leibniz, maintained a role for God, while others, like the reason-focused versions developed in the eighteenth century by Immanuel Kant and in the twentieth century by John Rawls, are at the heart of much secular thinking about nature today.

To sum up, our first understanding of nature is built around the idea of perfectibility. By this perfectionist way of thinking, while humans are not always perfect they will always strive towards perfection because nature is ultimately perfect and is therefore always, of necessity, seeking to return that which it has produced to perfection. Natural law for this position is the law that helps nature to perfect humans and their societies; while humans often stray from the path of perfection, natural law will bring them back to that path. In Aquinas's seminal version, God is 'the divine mind or reason from whose creative intellection emanate the essences or "natures" of all things … thereby constituting the *lex aeterna* or eternal law of the cosmos' (Hunter 2010: 477). Humans are able to know this law because, for Aquinas, as we indicated earlier, human nature 'shares the rational nature of God' (Hunter 2010: 477). This tradition is the basis of the most enduring form of opposition to legal-political government, usually featuring the idea that reason-based morality and/or religion are more powerful than law and politics, that law and politics have their place, but must know that their place is below morality and/or religion.

The rival to this understanding, which also has its roots in ancient Greek thought, this time in Epicurean and Stoic thought, refuses the idea of perfection. For Epicurus and his followers, as for the Stoics, humans are not perfect; their disquietude and constant violence make this empirically obvious, or so this understanding has it. More than this, nature is not perfect and instead of seeking perfection we should learn to accept nature's imperfect ways (including imperfect humans), we should study them carefully, and we should develop techniques to help us avoid the dangers of nature, especially the dangers of the uncontrolled passions of fractious humans.

In the face of the extremely violent civil wars in Europe in the sixteenth and seventeenth centuries, born mainly from religious turmoil, it is not surprising that these anti-perfectionist ideas were rejuvenated by some key early modern thinkers into something approaching a new science of humans and their society.[3] For example, in the sixteenth century they were rejuvenated by Jean Bodin in France and by Hugo Grotius and Justus Lipsius in the Netherlands (with not a little influence of thinking from earlier that century by Niccolo Machiavelli, writing in the different context of the political machinations of the northern Italian city-states), but their most famous advocates were the seventeenth-century political philosopher Thomas Hobbes in England and his main German follower Samuel Pufendorf. This early modern rejuvenation, particularly in Hobbes's hands, has become so influential as a mode of thinking about law and politics that, notwithstanding its roots in the ancient world, the seventeenth century is often reasonably treated as the beginning of the anti-perfectionist tradition, usually known as the civil tradition. This tradition is the basis of legal-political government. In the twentieth and twenty-first centuries the main Anglophone politico-legal inheritors of this tradition include Stephen Holmes, Ian Hunter, and Martin Loughlin (to name just a few).

In pursuing its three main goals – accept nature's imperfect ways (including imperfect humans); study these ways of nature carefully; and develop techniques to help avoid the dangers of nature, especially the dangers of the uncontrolled passions of fractious humans – the anti-perfectionist thinkers came to believe that while humans are fractious, they also, by their nature, fear death and its consequences and, because of this, ultimately crave peace. On the basis of this belief, the main advocates of this position, especially Hobbes, following Bodin's lead, proposed that the only sustained way in which humans' destructiveness can be overcome is by the strongest possible authority, which Hobbes called 'Leviathan' but which came to be widely known as the sovereign. Humans, by their nature, will fear and respect this authority, as the representative of God on earth and/or the repository of nature's power of life and death over them, and they will respect the peace this authority delivers to them in forcing them to control their formerly uncontrolled passions.

The rival anti-perfectionist understanding of nature, then, focuses on dealing with the difficulties thrown up by nature's immense power, including the immense power of humans. In particular this understanding has it that nature provides a means by which humans' natural capacity for destruction can, to a great extent, be curtailed. Natural law for this position is the combination of, on the one hand, the law imposed by the strongest earthly authority, which is itself seen to be natural, and, on the other, the law provided directly by nature, which determines that humans, because of their nature, will fear the sovereign and thereby respect the sovereign's laws, especially because those laws will severely punish those who fail to control passions and commit violence against others on the basis of this failure.

The rivalry between the perfectionist opposition to legal-political

3 In the third volume of his 'Visions of Politics' collection, Quentin Skinner, focusing on Hobbes, calls the new science of humans and their society 'civil science' (Skinner 2002b).

government and anti-perfectionist legal-political government itself is, as we keep saying, ongoing. The perfectionist position holds that the earthly laws of earthly rulers and the earthly politics in which they engage must always be subservient to externally sourced natural law. In this way, earthly law and earthly politics can and should always be judged against the externally sourced moral and/or religious criteria provided by natural law. In this way, if earthly law and earthly politics are not seeking perfection they are failing, and any force seeking to uphold natural law, which is by its nature superior, should strive to overcome regimes which are practising such inferior law and inferior politics.

For the anti-perfectionists, on the other hand, there can be no force superior to the sovereign. Any ruler who or which attains sovereignty, whether individual (a king or queen or prince) or assembly (a parliament), is, by the anti-perfectionists' understanding of natural law, the appropriate source of earthly law and politics, though only if he, she, or it is seeking to use his, her, or its rule to limit the dangers posed by uncontrolled passions.

Before we move on we should remind the reader that, as is the way with ideal-type accounts, not every position in the history of the development of legal-political government fits neatly within one or the other of our two rivals. While the figures mentioned above (Aristotle, Aquinas, Kant, et al. on one side; Hobbes, Pufendorf, et al. on the other) are indeed emblematic of the positions we have ascribed to them, many other important thinkers in the history of natural law and the history of the development of legal-political government are not. For example, thinkers like Alberico Gentili, Hugo Grotius, and Emer de Vattel, to name three, blend aspects of both traditions into their thought. Similarly, certain notions which might feature in any history of the development of legal-political government are blends of the two rival understandings of government at the heart of our narrative. For example, the notion of society can be said to be that which perfect nature provides, inasmuch as this understanding of nature includes humans as creatures too weak to survive without the companionship of other humans, and it can be said to be that which is achieved only when sovereign rule is strong enough to guarantee a peace so sustainable as to allow humans to interact without fear of death from other humans.

With this caveat in mind we trust the reader can see how present-day thinking about environment and climate change (be it legal, political, aesthetic, moral, religious, scientific, or economic thinking, or some combination of them) is influenced by both the lines of thought being laid out (even if the influence is indirect).

We move now to elaborate the rival visions of law and politics as they are understood, on the one hand, by perfectionist critics of legal-political government and, on the other, by the anti-perfectionist advocates of legal-political government. For the perfectionists, the appropriate sphere of operation for law and politics is the sphere of perfect justice, which nature extends to the entire planet and the entire universe. In this way, national governments – trapped in constant contests with one another – cannot be the highest authority. Inasmuch as the sphere of perfect natural justice is universal, a universal standard of law and politics must, by nature, override any national standards. So international law and politics are, or should be, superior, because by their nature they should strive towards perfect natural justice, being above the petty contests that define national law and national politics. In this way, universal perfect natural justice sets the standard for all law and all politics and, because they are refined reflections of this perfect standard, international law and politics are the most appropriate earthly vessels to pursue ultimate justice.

For the anti-perfectionists, or civil thinkers, of course, there is no such thing as the sphere of perfect justice. The places where law and politics operate are the places on earth where they have come to operate over time, in the particular territories that have come to be governed by law and politics. As such, the sphere of operation for legal-political government in the modern world has come to be the system of modern states initially established by the 1648 Treaty of Westphalia, which helped bring to a close the devastating Thirty Years' War in Germany (another of the religion-inspired civil wars referred to earlier). This system – which developed in piecemeal fashion as the rulers of various territories learned what legal-political government could and could not do in the wake of the civil wars – formally recognizes as a modern state any territory which is governed in line with the principles of sovereign rule set out above (the rulers must be the strongest force in the territory, able to dominate rival forces to the point of governing in the name of restricting the dangers of uncontrolled passions). Under this system the ultimate bearer of sovereignty in each territory came to be the state itself (which Hobbes referred to as the commonwealth or *civitas*).

In line with this, the state is, for the anti-perfectionist or civil way of thinking, the ultimate actor of legal-political government. International law and international politics are not superior, nor are they to be judged against a universal criterion of perfect justice. Instead international law and politics reflect the earthly relations between states, subject to all the politicking, diplomacy, and warfare which history suggests they will be subject to. Far from being about perfect justice, then, international law and politics are about treaties, alliances, and other such markers of earthly interactions. For this position, national laws and national politics are neither necessarily inferior nor necessarily superior to international law and politics. As nations are, at least since the middle of the seventeenth century, usually defined as states, national laws and politics for some states are sometimes able to override the international obligations of those states, while for other states their national laws and politics will be forced into submission by international law and politics (which in practice often means 'in the name of stronger states involved in the particular treaty or other such arrangement').[4]

---

4 Obviously there are many exceptions to our point about nations being 'usually defined as states', including instances in which suppressed national sentiments within existing states become strong enough that those states are broken into smaller states, thereby legitimizing the previously suppressed claims to nationhood.

What this means for the legal-political government of the environment and climate change

With the rival understandings of the basis of government laid out, we need to say something more by way of introducing each understanding's treatment of environment and climate change, in preparation for the detailed arguments to be presented in later chapters. Of course, as objects of government, environment and climate change are not ideal types in the way that our two rival understandings are ideal types; instead, they are real-world, real-time entities. As such, they do not always fit neatly within one or other of our two rivals. It might be said that certain extreme environmentalist positions in present-day debates about climate change, for example, are entirely the product of the perfectionist camp, which mainly opposes legal-political government in the name of morality and/or religion, while treaties arrived at after laborious international negotiations, such as the Kyoto Protocol, are entirely the product of the civil or anti-perfectionist advocates of legal-political government. But we think it wise to remain cognizant of the fact that all the contributions to these debates are real-world positions and most will contain at least some traces of both the rivals we have been at pains to describe.

In line with this point, we might say that most present-day debates about the environment and climate change have a long history but a short memory. They have a long history in that, as we just said, it would be very difficult for a contribution to be made to these debates that is not sourced, to one degree or another, in these rival camps. Yet they have a short memory inasmuch as there is precious little awareness displayed of this long history. Too many contributors to the debates speak or write as if their particular vision of the environment and global climate change is a-historical, as if their thinking is informed by present-day facts alone.

To help make our 'long history' argument clearer, we will extend the above distinction between a hypothetical extreme environmentalist position which opposes compromise and a hypothetical position concerned to negotiate workable treaties. This will give us a chance to discuss the ways in which aesthetic, moral, religious, scientific, and economic discourses feed into our story, which so far has focused mostly on the history of legal and political discourses.

The extreme environmentalist position set out above might involve an aesthetic attachment to the beauty of the environment born of, for example, nineteenth-century Romantic movements in literature and art, movements which were often committed to a perfectionist view of the environment, often as the product of a perfect God, or a perfect Nature, or both. But it might equally involve an aesthetic attachment to the environment born of nothing more than a personal calculation that big expanses of water and old forests are better to look at and listen to than urban decay. There is no necessary connection between a given aesthetic approach to the environment and a perfectionist understanding of nature or an anti-perfectionist understanding, though of course it would help an analysis of any given debate about the environment to seek to determine which aesthetic discourses are being drawn upon.

With regard to the economy, the extreme environmentalist position might involve a perfectionist antagonism to any economic development of the environment. But it might equally involve a belief that certain types of economic development can enhance human interaction with the environment at the same time as allowing certain firms and individuals to make a monetary profit from it. There is no necessary connection between a given economic approach to the environment and an anti-perfectionist understanding of nature or a perfectionist understanding, though again it would help an analysis of any given debate about the environment to seek to determine which economic discourses are being drawn upon. And if we combine these two elements, it might be that a firm seeking permission to mine for shale-based gas in a wilderness area is only interested in profit and will always seek ways to minimize the impact of legal and political hurdles, as is consistent with a certain type of thinking about the economy as an independent domain, a domain which emerged only in the nineteenth century (Walter 2008a, 2008b, 2011). But it might equally be that the principals of the firm have a personal aesthetic attachment to the beauty of the wilderness they are keen to mine, perhaps born of the nineteenth-century Romantic movements mentioned above, and yet, in being no less attached to their firm, also think it possible to mine the area without harming its beauty. In this scenario, the firm will be willing to not only follow legal and political directives but even to hire in specialists of their own to help protect the beauty of the area.

The reader, we feel sure, is getting the picture: the different discourses of the environment – legal, political, aesthetic, moral, religious, scientific, and/ or economic – can combine in any number of ways, some completely in line with legal-political government, some completely in line with opposition to it, and some located between the two. We need say no more about combinations of legal, political, aesthetic, and economic discourses of the environment. But we still need to say something about the diverse character of the moral, religious, and scientific discourses which are nearly always present in modern debates about the environment.

The intersections between the environment and each of moral, religious, and scientific discourses are perhaps more one-sided than are the intersections of the environment and the legal, political, aesthetic, and economic discourses discussed so far, but they are not by any means completely one-sided. The intersection between the environment and science is easier to explain than those between environment and morality or religion, so we will tackle it first. It is easier to explain simply because a battle between a perfectionist view of science and an anti-perfectionist view has been waged very publicly over the last ten years or so. This battle is extremely germane to our book; it is that waged over the role of science in determining and possibly mitigating anthropogenic climate change.

In this setting, a view of science as the highest possible authority on whether anthropogenic climate change is occurring has seen it afforded a quasi-perfectionist reputation, though rarely has this been done by scientists themselves. This 'highest possible authority' view has been opposed on two sides, as it were. On the one hand, albeit only occasionally, it has been opposed by an ultra-perfectionist view which regrets the fact that science is not doing enough to demonstrate the 'ultimate truth' of climate change. On the other hand, and much more often, it has been opposed by the view, shared by most scientists themselves, that science is not about the search for perfect knowledge but is, rather, concerned to establish only the best possible knowledge available at a given time by testing the evidence at hand. In the midst of this complex ongoing argument about science and its role, a plethora of opinions has emerged about whether climate change has ever taken place, about whether it is now taking place, about whether it will keep taking place, and so on.

So, we can conclude that at the intersection of the environment and science, as with the other real-world intersections we have discussed, perfectionist and anti-perfectionist understandings cannot be neatly separated one from another in the way that they can in our ideal-type account.

Moral and religious discourses may seem at first glance to be completely dominated by perfectionist thinking, inasmuch as they are discourses concerned with ultimate standards for living and dying. But, yet again, in the real world this is not always the case. In the course of their intersection with the environment, moral and religious discourses are often open to anti-perfectionist thinking. To explain what we mean it is best that we deal with religious discourses and then link them to moral discourses. Probably the most obvious examples in the long period at the centre of our concerns (the sixteenth century to the present) are the many subtle varieties of religious doctrine developed as part of and in the wake of the confessional disputes/wars in Europe often known as the Reformation. In this way the perfectionist thinking of strict Roman Catholic doctrine (and strict Lutheran doctrine for that matter) was challenged by different currents of anti-perfectionist thinking behind the many forms of Calvinist and other dissenting religious doctrines that emerged over hundreds of years. Some, such as the anti-perfectionist arguments of Christian Thomasius, a student of Pufendorf's, were very well developed:

> [A]ccording to Thomasius's anti-doctrinal, Epicurean style of Protestantism, the number of divine commandments relevant to salvation could be reduced to just three: to love God, and one's neighbor, and to have contempt for oneself (as a creature of passions always prone to disorder). As a result, all of the things that the competing confessions declared to be essential, and over which so much blood had been spilled – all of the church liturgies and sacraments, the vehement doctrinal disputes over the Trinity, the nature of Christ's presence in the Eucharist, the relation between Christ's 'two natures and one person', and so on – could be declared to be matters of moral indifference, turning them into matters of 'Christian freedom' or else of political regulation.
>
> (Hunter et al. 2007: xvi–xvii)

This is to say that Thomasius kept separate 'the fields of religion and politics in a way that … removed salvation from the domain of church ritual … and removed political authority from the domain of salvation'. The effect of this move was to leave 'individuals privately free to pursue salvation as they saw fit', and to leave the churches 'with the status of voluntary associations under state supervision' (Hunter et al. 2007: xvii).

In the domain of morality the influence of the alliance between Epicureanism and dissenting forms of Protestantism was equally important. The idea that there could only ever be one true morality, by which all things must be judged, was challenged in a variety of ways. For instance, the conventionalism developed by those, like Hobbes, keen to build a science of human behaviour held that all morals are in fact historically developed conventions. It should be noted that for this position morality was still vital, as a guide for living and dying, but it could no longer be assumed to be the same at all times and in all places (this style of commitment to morality remains important to anti-perfectionist arguments today). In a very similar vein, some international law thinkers, such as Vattel, developed a flexible way of understanding the moral obligations of parties to international treaties. This mode of understanding, in seeking 'to show how general principles can be explicated to suit particular cases and circumstances', is often known as casuistry. Casuistry is 'dedicated to managing the circumstances where the principles do not apply … by mobilising arrays of lower-level conventions and customs that have been developed for and through specific "cases"' (Hunter 2012: 14).

The key question and the structure of the book

We have said enough to allow us to now pose what we regard as our book's key question: how do law and politics, entwined in various complex relations with sovereignty, the state, morality, religion, economy, aesthetics, and science, attempt to govern the environment and climate change?

We should emphasize, before we discuss the book's structure, that in answering the key question we will not attempt to hide our qualified admiration for the system of legal-political government. We say 'qualified admiration' because its record, as we will say many times hereafter, is checkered. As well as delivering the benefits we will discuss in the coming chapters, this system has been involved in colonial excesses, the French Revolution, the American Civil War, the First and Second World Wars, many economic depressions, to name just some of its more obvious inglorious moments. More than this, it quite often promises much more than it can deliver on a number of its noble goals, like wiping out poverty, improving the living conditions of minorities, generally looking out for the downtrodden, and, particularly relevant to us, helping those affected by natural disasters, such as Hurricane Katrina or the more recent Japanese earthquake and tsunami.

As a general remark on the other side of the ledger we might point to the system's resilience in at least always trying (or nearly always trying) to face up to its failures and do something about them. In what we will later describe as its 'trial and error' mode, legal-political government usually tries to fix its failures with still more legal-political government, changing a strategy here and a tactic there, using different personnel, and so on. The results may be only marginally better, and sometimes worse (on this, see Prins and Rayner 2007: 973), but it usually keeps trying until it meets with at least a modicum of success. Perhaps, in this regard, the system is worthy of an adaptation of Winston Churchill's famous assessment of democracy: 'Democracy is the worst form of government, except for all those other forms that have been tried from time to time' (Churchill 1947). We leave that for the reader to judge.

What, then, will each of the other six chapters contribute to our answer to the book's key question?

Chapter 2 will focus on the sixteenth and seventeenth centuries (though it will occasionally look in on the eighteenth) in building a more layered picture of the new form of government we are calling legal-political government. This chapter will include a number of examples of the way this form of government is used to govern the environment and climate change, but it will be left to later chapters to present our detailed account of this matter.

Chapter 3, starting in the eighteenth century, but focusing mostly on the nineteenth and twentieth, will explore the way in which the environment emerged through a combination of aesthetics and science as an object of legal-political government.

Chapters 4 and 5 will both seek to explain the role of law in present-day legal-political attempts to govern the environment. Chapter 4 will be more micro-focused, dealing primarily with the private common-law government of the environment, in particular a form of it called toxic tort, though only after it has situated law's dealings with environment and climate change in a particular 'sphere of operation'. In this chapter we will also introduce the argument that the legal-political government of the environment and climate change has a lot to do with governing risk. Chapter 5 will be macro-focused at the level of the nation state, inasmuch as it will deal primarily with the public-law government of the environment through legislation, something which sees public law combine with public interests.

Chapter 6 will be, as it were, global-focused, dealing with the ways in which legal-political government is currently seeking to govern climate change by using a range of strategies, some local, some national, and some international.

In our conclusion, Chapter 7, we will summarize our main themes and link these to a discussion of two brief case studies of the legal-political government of the environment and climate change in Australia. These case studies offer a particularly vivid picture of the different mixtures of politics, law, sovereignty, the state, the economy, science, aesthetics, and morality which are used by legal-political governments as they go about their governing.

2 Legal-political government as a new form of government

Introduction

Over the course of the sixteenth and seventeenth centuries legal-political government emerged in Europe as a new mode of ruling. It spread across many other parts of the world in the centuries that followed, often replacing modes of rule based on religious belief systems or on force alone or on some combination of them. The new form of government both relied upon and contributed to a revived anti-perfectionist understanding of humans which had its source in the ancient world. From the late eighteenth and into the nineteenth century the new type of government slowly became the dominant way of governing what we now call the environment, though this could not happen fully until the twentieth century, which was when the environment first became an entity in its own right. We will focus exclusively on the emergence of the environment as an object of legal-political government in Chapter 3, but in the present chapter we have a different task.

Here we need to expand upon our introductory discussions of the key components of the new mode of government – law, politics, sovereignty, the state, the economy (as a separate domain), morality, and religion – and we need to introduce a component which we did not mention in Chapter 1, the notion of interests. In what follows – woven through the separate sections on each of law, politics, sovereignty, the state, the economy, and interests – we will describe the ways in which legal-political government, in giving pre-eminence to its central governing mechanism, the sovereign state, developed its own anti-perfectionist roles for the various forms of morality and religion that continue to operate in the territories it rules. In doing this, we cannot stress enough, legal-political government does not itself act in accordance with the strictures of these forms of morality and religion. Instead, it acts in accordance with its own normative trajectory. This trajectory attempts to satisfy just one overriding norm, a norm which emerged in the second half of the seventeenth century when the early legal-political governments (in England, France, Germany, and the Netherlands) succeeded to such a degree in their quest to develop techniques to help the subjects of their rule avoid the dangers of their own uncontrolled passions that the subjects themselves began to appreciate the growing spread of peace, security, well-being, and prosperity this type of government was delivering to them. In other words, they began to appreciate that the power of their governments was – to anticipate a formulation we will come to later in this chapter – a positive *power to*, not just a negative *power over*. In line with this, the sole norm of legal-political governments is: the pursuit of the widest possible appreciable spread of peace, security, well-being, and prosperity among the humans within each territory being governed (this norm is sometimes referred to simply by the term 'civil peace', a nomenclature we will often employ from now on when referring to the normative direction of legal-political government).

In saying that the legal-political system of government is distinguished from other systems by its *appreciable* commitment to the widest possible spread of peace, security, well-being, and prosperity among the humans in each territory being governed, we are highlighting the fact, discussed briefly in Chapter 1, that in legal-political government the law must be strong enough to temper politics in such a way that they work together, however easily or uneasily. If this is not the case, if politics is able to operate only or mainly on the basis of what we will describe later in this chapter as its own raw power, government is still possible, of course, but not legal-political government. If a government operating within such a politics-dominated system offers a commitment to peace, security, well-being, and prosperity, it will definitely not involve delivering individual freedoms to the great majority of the population it is ruling (raw politics will prevent it from doing so, as we will explain later). It is possible for it to deliver such freedoms only to a minority, possibly only to the ruling elite. But if, on the other hand, the government of a law-and-politics country – in which law, by definition, is strong enough to work alongside politics as an equal or near-equal force – offers the same commitment, it will be an *appreciable* commitment. This means that the great majority of the population of that country will have the opportunity to appreciate and exercise the individual freedoms that the genuine maintenance of peace, security, well-being, and prosperity will have delivered to them.[1]

All this is to say that we use the word 'appreciable' to mark the fact that the great majority of the populations of law-and-politics countries are *in a position to appreciate and exercise* the individual freedoms which are made available to them by the fact that their legal-political government is able to maintain the country's commitment to security, well-being, and prosperity.[2] Of course

---

1 We are using the term 'law-and-politics country', we trust it is clear, to refer to countries ruled by a legal-political government. But we know this is not the usual term for such countries. A more usual practice is to stress the fact that they are countries in which the great majority of people can enjoy meaningful individual freedoms by calling them 'free countries', or 'part of the free world', or perhaps 'countries with a liberal system of rule'.

2 Later in the chapter we will discuss the fact that on rare occasions – only those when a law-and-politics country is in danger of being wiped out by enemy attack, either from within or without, or by some other such calamity – these freedoms can be temporarily curtailed by legal-political governments.

some members of such populations may not want to exercise these freedoms, for whatever reason – possibly for religious reasons, possibly for cultural reasons, possibly for some other reason. Under legal-political government they are free to refuse the freedoms, though only as long as they do not then threaten the rights of others to exercise them, which would be threatening the civil peace.

Another important thing to note about legal-political government's normative dimension is that while it allows an important role for other moral codes or conventions, as we explained in the previous chapter, its overriding norm (civil peace, or the widest possible appreciable spread of peace, security, well-being, and prosperity among the humans within the territory being governed) is ultimately more important than any of these other norms (and we mean 'ultimately more important' here in its gravest sense – 'ultimate' is when the state itself is threatened with annihilation; something which will be discussed in more detail later; for now it is enough to say that in any law-and-politics country no norm other than the maintenance and enhancement of appreciable civil peace can ever legitimately be used to judge law or politics). In any given law-and-politics country, the civil peace norm is not always easy to discern, especially because it is often seen to be at odds with other moral/religious norms of different groupings within that country (such as protest groups, business groups, churches, and some media outlets) and/or at odds with those of other countries (which are likely to share the same broad norm of civil peace within their territories but interpret it differently). The competing norms and competing interpretations of the overriding norm came to be expressed as 'interests', whether those of the country in question or those of other groups within or across countries (which is why we will say more about the importance of interests later in the chapter).

And still one more important thing to remember about the system's one norm is the sizable caveat we issued not that long ago: in pursuing its sole norm, legal-political government has a long record of serious failures. While it has proved itself resilient, its capacity to deliver on its norm is not limitless; indeed, in some places and at some times it is severely limited.

When we have completed our treatment of each of law, politics, sovereignty, the state, the economy, and interests, our final main section will explore the complex ways in which these various elements combine into a single force, legal-political government. As our link between all the material covered in this chapter and the material to be covered in the other chapters, we will be sure to offer regular examples of environmental government.[3] In

---

3 We hasten to add, apropos these examples, that our strong focus on the period 1500 to the present does not mean we think the environment was totally ungoverned before 1500. Inasmuch as English usage allows broad meanings of the words 'govern' and 'environment', as well as the more particular meanings at the heart of our argument, one can sensibly propose that those farmers in the ancient world who learned the benefits of terracing by trial and error were governing what we now call the environment, as were shepherds who used trial and error to develop increasingly efficient techniques for husbanding their sheep through different seasonal conditions, and so on. We will deal with older modes of governing what we now call the environment in more detail in Chapter 3.

the chapter's conclusion, as well as consolidating our main points, we will discuss the separate roles legal-political government has given to newer entities, specifically science, and the new role it has developed for older entities, specifically aesthetics.

Before proceeding we wish to reiterate that legal-political government, in both relying upon and contributing to a new anti-perfectionist understanding of humans, leans towards an anti-perfectionist understanding of the goals and aspirations of governing on a large scale. By using the formulation 'leans towards' we are conceding that while this form of government mostly avoids perfectionist thinking about that which it attempts to govern, it can never be expected to do so totally. The world is too complex for that. Legal-political government, we might say, while always leaning towards anti-perfectionism, is forever dealing with continual contests between perfectionist and anti-perfectionist thinking. But we do not want to take the concession too far. In the spirit of not taking it too far we wish here to emphasize more forcefully than we have done previously what legal-political government is *not*: legal-political government is not government by religion-and-morality, it is not government by private moral conscience, and it is not government by lofty ideals.

Law for legal-political government

We could formally define law for legal-political government as an ensemble of juridical institutions operating inside the state. However, as this definition might be regarded as circular, we need to strengthen it. We can do so by drawing on some of the insights of Martin Loughlin. For Loughlin, law is both the constitution and the set of laws operating within its realm. This definition might at first glance look just as limited as the 'ensemble of juridical institutions' definition, but it is not. Loughlin does not mean the constitution as a document, written or unwritten, which might be treated as a sacred object. Instead he has something more active in mind – the actual constituting of the sovereign as the authority that issues all other laws. The constitution, as the constituting law, this is to say, is the legal authority that was 'created by the secularization, historicization and positivization of natural law' (Loughlin 2009: 5) (a process conducted, as we saw in the previous chapter, mostly in the sixteenth and seventeenth centuries, though of course it can never be said to be complete, for the authority of the sovereign has to be always maintained).[4] When this foundation is in place, Loughlin continues

---

4 In basing his account of the law on the authority of the sovereign, as we will see across the course of this chapter, Loughlin is very much a Hobbesian. Hobbes says that the sovereign is: the sole entity with the power to ensure that covenants bind the parties to them (1845b: 157), such that property is guaranteed (1845b: 165, 233; 1845c: 131); the sole entity with the power to make the laws of nature binding (1845b: 253); the sole entity with the power to interpret the law (1845b: 262–63); and in all this the sovereign must be the sole legislator for any commonwealth (1845b: 252, 336; 1845a: 77; 1845c: 131).

(2009: 5), all other laws are 'created as a consequence of' the constituting law; these 'other laws' are often collectively called 'positive law'.

In the terms of Loughlin's definition, the legal-political government of the environment and climate change in the twenty-first century relies on the continuing success of the constituting law in guaranteeing the authority of the sovereign state, which in turn guarantees the greatest possible appreciable civil peace among the humans within the state's defined territory. As long as the constituting law continues to be successful in this task the various positive laws which are aimed at environmental protection (more on these in Chapter 5), as well as various private law dealings with the environment (such as tort; more on this in Chapter 4), have a solid base on which to work. In this way, while attempts by governments or agencies of government to rid water supplies of dangerous pollutants, for instance, date back to ancient times, their effectiveness was always limited by the limited reach of older forms of authority; twentieth and twenty-first century legislation and its enforcement have a much greater chance of being effective because they are part and parcel of the far-reaching sovereignty of any law-and-politics state (of course one should take this 'much greater chance of being effective' as meaning no more than it says; it is not, and can never be, a guarantee of unpolluted water).

Loughlin's succinct account of law in what we call legal-political government helps to make sense of David Saunders's (1997: i–viii) more combative definition of law for this type of government – combative because Saunders is determined to confront the ongoing campaign by moral and/or religious perfectionists to undermine the constituting law by challenging the sovereign's capacity to restrict the role of morality and religion (something sovereign authorities themselves have been forced to confront since the early modern period, lest this campaign threaten the civil peace that legal-political government works so hard to maintain). Saunders wants the law component of what we call legal-political government to be defined within the terms of its early modern role of imposing peace upon competing moral and religious communities, especially in the light of the fact that most of these communities displayed very little willingness to tolerate each other, being all too keen to 'resolve' issues by wiping out all members of competing communities.

For Saunders, this is to say, despite the fact that the project of legal-political government gained more and more ground in the wake of its success in bringing one hundred and fifty years of European civil wars to an end, moral and religious perfectionists never stopped fighting back against the idea that law is a form of ordering civil life which relegates religion to the private sphere and which operates with a norm concerned only with fostering a widespread and appreciable civil peace. The mammoth task of separating law from traditional morality and public religion, he argues, remains incomplete. The moral and religious perfectionists (a camp into which Saunders places not only religious fundamentalists but also many modern 'critical' intellectuals) persist in seeking to realign both the conduct of government and the conduct of the legal apparatus with supposedly universal moral principles (Saunders

1997: i). Referring to the law component of legal-political government as 'an exceptional accomplishment' in a period in Europe when 'a proselytising religious culture' was dominant, Saunders worries that today this approach to law, despite its achievements, is 'our own unfinished business', that 'separation of spiritual discipline from secular government and conscience from law was never complete'. In other words, he laments the 'incessant claims of critical intellectuals to reshape governmental institutions and the legal apparatus in accordance with ... [an eternal, universal] moral principle, typically some vision of individual autonomy or communitarian self-determination' (Saunders 1997: viii). These intellectuals, he says, should recognize that without the state and the law, which they seem to detest, they would have 'no secure platform from which to project their vision of a new society and to preach their faith in redemptive moral politics' (Saunders 1997: 9). Driven by the goal of 'a moral society beyond the reach of the State', a society grounded 'in inalienable rights and fundamental freedoms', these critics all too easily forget the role of the state, the law, and politics 'as the grantor and protector of religious freedom' and see them instead as 'the great threat to freedom' (Saunders 1997: 9–10). In this, such thinkers have 'reoriented themselves to a future society where morality would again govern and where men, escaping the confines of coercive legal citizenry, would at last be freely themselves' (Saunders 1997: 10).

This discussion should help make clear why we insist on including within a discussion of law's role in legal-political government a discussion of morality and religion. We think it impossible to understand law as a component of legal-political government without understanding its insistence that it has created for itself a different type of moral compass, one which owes no allegiance to a particular religion. This is easier to comprehend in the light of claims made by the legal-political version of law that it has a different type of conscience than that called on by its perfectionist critics. Before we turn to this different conscience, an example is called for.

In any legal jurisdiction within a law-and-politics country a proposal to build a dam is certain to create a number of disputes. The law will be called on to help resolve disputes over where the dam might best be located (disputes likely to involve affected landholders, mining leaseholders, nearby residents, environmental lobby groups, and industry lobby groups, among others). It might also be called on to help resolve disputes about whether the proposed dam is in accordance with the country's international treaty obligations concerning protection of certain types of flora and fauna, for instance, and of course it will be called on to help set up contractual obligations between the parties involved in building the dam and to mediate grievances between the contracted parties. In all this, the law component of legal-political government will be effectively operating to promote as wide an appreciable spread of civil peace as possible *without* recourse to an external morality (external, that is, to legal-political government). This does not mean of course that it will be operating immorally or amorally. Rather, it means that law's morality is concerned with the public promotion of widespread appreciable civil peace, not with the pursuit of supposedly higher goals to do with the perfectibility of humans. In this, law, as it operates in legal-political government, answers to a public conscience and not to a private conscience driven by supposedly universal moral and/or religious standards. As Hobbes puts it, 'the law is the public conscience' (1845b: 311).

If the idea of a public legal conscience sounds odd, this may be because, as Saunders argues, 'the individual moral conscience' is heavily promoted 'as the ultimate site of an uncompromising universal adjudication'. For those who promote 'the individual moral conscience', the goal of 'a moral society' seems always to trump the seemingly mundane goal of civil peace pursued by 'the administrative State'. Such perfectionist thinkers are driven by 'a constant moral dissatisfaction with existing institutions', institutions which, in their eyes, are antithetical to 'a future moral society' (Saunders 1997: 10–11).

Saunders argues that while a private moral-religious conscience is 'regulated by the "truth of a creed"', the 'public legal conscience' is 'regulated by the "laws of the land"', being 'strictly "civil and political", namely pacific and prudent' (Saunders 1997: 22, quoting Thomas). He traces the history of the public legal conscience to the sixteenth and seventeenth centuries, a time when,

> 'it was generally believed that [private religious] conscience, not force of habit or self-interest, was what held together the social and political order … Every attempt by the State to prescribe the forms of religious doctrine and worship tested the consciences of those who believed it was their duty to obey the laws of the land but were also persuaded of the truth of a rival creed'.
>
> (Saunders 1997: 21–22, quoting Thomas)

Saunders focuses especially on the significance of two 'dialogues' published in 1528 and 1530 by Christopher St German, 'a barrister of the Middle Temple in London', which sought '"to demonstrate that the common law rather than the decrees of the church should govern the consciences of Englishmen"' (Saunders 1997: 23–24, quoting Yale). St German's basic argument eventually became widely accepted in common-law circles and by the end of the sixteenth century figures such as Selden and Hale, among others, were arguing forcefully about the need to keep religious conscience out of the law (Saunders 1997: 25–26). The court of equity, in particular, fostered the basic idea of a legal conscience: 'This self-limiting conscience of the court is the law's positivity, its delimitation of its jurisdiction and its objects of administration' (Saunders 1997: 26). Where 'Christian enthusiasms … sought to impose confessional conscience across the whole of life' this very different type of conscience helped courts 'to adjudicate legally rather than confessionally' (Saunders 1997: 27).

The reader will see from this discussion that as legal-political government emerged it did not seek to obliterate those religions and those moral codes which claimed to be timeless and universal, and nor does it seek to do so today. Instead it seeks to move the religions and moral codes which claim to be timeless and universal from the public sphere into the private sphere (as it has done since the seventeenth century), replacing them in the public sphere with a public civic morality based on law. Where the religions and moral codes which claim to be timeless and universal focus on the possibility of perfecting humans, often in the image of a perfect God, legal-political government focuses on the idea that because humans are not perfectible the primary task of rule is to use law and politics to actively limit the harm humans would do to one another without such intervention (as they kept proving over and over in early modern Europe and, sadly, as they keep proving in too many parts of the world today). As such, governing the environment and climate change for this type of government is not fundamentally about preserving the supposed perfection of nature, it is about regulating the interaction between humans and nature in such a way as to allow humans to gain long-term benefit from nature towards the goal of widespread appreciable civil peace *without* harming the chances of continuing to do so in the future. To put this point another way, governing the environment and climate change for this type of government is, at any given time, about trying to control the ever shifting balance between the benefit of humans (towards as wide an appreciable spread of civil peace as possible) and the maintenance of nature – an imperfect endeavour if ever there were one.

In this way, legal-political government does not reject morality as a guiding force so much as reject the idea that morality is timeless and universal. It seeks to redefine morality in terms of real-world outcomes of public government *in particular places at particular times* (towards the goal of greater civil peace), as something, that is, which has its own history. Blandine Kriegel (1995) explores this history in some detail. She takes her lead from Nietzsche in arguing that when 'the Greco-Roman heritage lost its early appeal ... it was Judeo-Christianity that became the moral tutor of the West' (1995: 53). She does not, however, share Nietzsche's criticisms of Judeo-Christian morality; she is interested in particular Judeo-Christian propositions on how humans should behave and how they should be treated. More than this, she is interested in the ways in which these propositions were taken up by early modern thinkers in attempting to construct a new form of government around the sole norm of the pursuit of the widest possible appreciable spread of peace, security, well-being, and prosperity among the humans within each territory being governed. She is particularly interested in those early modern French and English theorists (Bodin and Hobbes especially) who developed from these propositions 'a doctrine of individual right', not towards the goal of individual liberty for its own sake, but for the sake of the emerging state's quest for an appreciably greater civil peace, a doctrine which focused on 'the relationship between, and the limits of, the rights of authorities and the rights of individuals' (1995: 33–34). Kriegel argues that the historical roots of this doctrine lie in both the Old and New Testaments. The Old

Testament stresses humans' obligations, in terms of their covenant with God, while the New Testament emphasizes the 'inalienable value' of each and every individual, something foreign to the ancient Greeks and Romans. The early modern legal and political thinkers she is keen to highlight 'obstinately and patiently established the foundations for personal security and liberty, those fundamental rights that enabled us to emancipate ourselves from the state of war and servitude, and which we today take for granted [as human rights]'. In line with this, the idea of 'the supreme dignity of the human being' led to 'the process by which slavery became indefensible'. The law component of legal-political government, she suggests, was infused with the idea within Judaism that law is a form of future-proofing: 'The law, in sum, transcends territory and defeat and the ephemeral lives of individuals; it assures, so long as it is safeguarded and transmitted, the perpetuation of an identity' (Kriegel 1995: 34–37).

Kriegel does not dismiss the importance of faith. For her, faith is central to the privatizing of religion that was part and parcel of the new style of governing, a development which eventually came to be called the separation of church and state: 'The modern state ... left to the individual and to the church the task of salvation and concerned itself with the operation of law and politics (1995: 37). As such, the 'juridical sphere ... enjoyed considerable expansion: collective law had been extended to areas where only fragmentary rights had been acknowledged; a general system had taken the place of piece-meal rules with limited application; political right, in sum, had emerged to overtake civil law' (1995: 61).

In concluding this section it is important that we acknowledge that the main form of law for legal-political government is one that we have not so far mentioned directly: public law. We have not mentioned it directly because we think it cannot sensibly be discussed until we have introduced the politics component of legal-political government. Public law is a pillar of legal-political government, in that, as well as being the source of legislation – the main mechanism used by legal-political government in the normal course of its work of directing the population it rules towards particular ends – it is the form of law that does most to restrain the raw power of politics such that politics can play its part in delivering widespread civil peace, instead of delivering constant deadly violence, of which it is quite capable, a matter to which we now turn.

Politics (and public law) for legal-political government

Our definition of politics acknowledges the seminal work of Carl Schmitt (1976), who conceptualizes politics in terms of relations between friend and enemy. In allowing these two fundamental categories, friend and enemy, to be always in flux, Schmitt encourages a focus on the necessarily contested nature of politics, particularly in recognizing that the friend–enemy relation is always potentially a violent, sometimes deadly relation. This is what gives politics its great force, its raw power. Politics is, as it were, the engine of all governmental projects. Of course there is more to politics – as an arm of legal-political government – than perpetual contest. There is also getting things done and attempting to get things done.

The 'getting things done' aspect of politics is well summarized by Loughlin (2003: 41) who, in focusing on activities such as administration, discussion, diplomacy, and policy formation, defines politics as that which 'enables the activity of governing to be effectively conducted'. He adds two points of clarification: (i) by this definition of politics, 'Political power maintains the public realm' (2003: 77); and (ii) this type of political power is 'power to' not 'power over' (2003: 158–59).

In times of widespread civil peace – times, that is, free from genuine threats to the existence of the state – what we are calling the 'getting things done or attempting to get things done' side of politics is by far the more active and obvious. But it might reasonably be said that what we are calling the 'friend-enemy' or 'raw power' side, including its threat of deadly violence, is the more dominant, even while it is in the background. As Max Weber puts it (and it should be remembered that Weber influenced Schmitt), '"It is absolutely essential for every political association to appeal to the naked violence of coercive means in the face of outsiders as well as in the face of internal enemies"' (Turner and Factor 1987: 343, quoting Weber).

Obviously government becomes impossible whenever and wherever the 'raw power' side of politics dominates the 'getting things done' side for any length of time. Down that path, as we keep saying, lies ongoing destructive violence. This is precisely where public law comes into its own. As Loughlin (2003: 1) puts it, 'the basic tasks of public law … can briefly be defined as those concerning the constitution, maintenance and regulation of governmental authority'. Public law was developed, this is to say, so that legal-political government could maintain a practical focus on the maintenance of 'security, liberty, and prosperity'. In this way, public law is the most thoroughly developed form of the 'constituting law' we outlined earlier:

> With the transmutation of the king's servants into officers of the state, a decisive step was taken in establishing an impersonal, specialized administrative apparatus that could exploit developments in printing, record-keeping, indexing, and such like.
>
> (Loughlin 2003: 8; see also Brewer 1989)

It is impossible to exaggerate the importance of the development of an 'impersonal, specialized administrative apparatus' to the development of legal-political government. We will say more about this in a later section, when we focus on the state. To complete our discussion of public law, we offer a sketch of the development in Germany in the late seventeenth and early eighteenth centuries of a type of public law which aimed specifically to strengthen the hand of the emerging state as it sought to move still further from the devastating violence of the Thirty Years' War, the war sparked by the long run of inter-confessional disputes, which, as we noted earlier, is sometimes known as the Reformation or, more accurately in this context, as the first and second reformations. During these troubled times Lutheranism broke with Roman Catholicism, and then Calvinism broke with both Roman Catholicism and Lutheranism. The Thirty Years' War was brought to an uneasy end by the 1648 Treaty of Westphalia, mentioned earlier, itself a part of the type of public law under consideration. Among the intellectuals behind this type of public law were two figures introduced earlier: Samuel Pufendorf, the German disciple of Hobbes, and Pufendorf's pupil Christian Thomasius, who went on to become a leading jurist in his own right.

These two thinkers worked tirelessly to help produce a new body of natural-law-as-public-law, one that might keep the church from interfering with the state's growing authority, a body of public law, that is, to support that form of rule which is restricted 'to the purely worldly domination of a territory' (legal-political rule). Hunter (2001: 63; see also: 76–77) tells us that in Thomasius's foreword to his 1707 translation of Hugo Grotius, he advocated a combination of '*jus publicum* and *Staatsrecht*'. In this way, Hunter continues, Thomasius sought to support the efforts of Pufendorf in developing 'the theologically indifferent *Staatskirchenrecht*', as a pointed form of public law:

> It was indeed through the protracted elaboration of the political-legal instruments required to deal with the religious civil war that German political or public law … gradually became independent of Roman law, employing the latter's categories for the scaffolding for these great works of legal construction, but filling them with contents suited to purposes unknown to the Roman legists.
>
> (Hunter 2001: 83)

In this, these and other jurists succeeded in detaching German public law ('*jus publicam*') 'from all higher-level moral and theological ends, thereby allowing it to be treated as a set of purely instrumental commands required to achieve social peace' (Hunter 2001: 83–84). As Saunders (2002: 2179) argues, these thinkers eventually helped defeat the idea that law and politics must form 'a stairway to salvation'.

To help cement in place our point about the way public law developed at the intersection of law and politics, it is worth saying a little more about Pufendorf's and Thomasius's anti-perfectionist approach to politics. Hunter describes their position as 'anti-metaphysical voluntarism' (one can see immediately the influence of Hobbes on their thinking): 'Pufendorf and Thomasius denied … the transcendent truths of man's moral nature or moral community', in favour of an understanding of 'his limitless capacity for mutual self-destruction' (2001: 89). Pufendorf, Hunter makes clear, knew what was required, offering a devastating picture of, natural man as a creature whose weakness … necessitates sociality for survival but whose 'vices render dealing with him risky and make great caution necessary to avoid receiving evil from him instead of good'. Unlike the beasts, man's appetites for sex and food are limitless and impossible to satisfy. Moreover: 'Many other passions and desires are found in the human race unknown to the beasts, as, greed for unnecessary possessions, avarice, desire of glory and surpassing others, envy, rivalry and intellectual strife. It is indicative that many of the wars by which the human race is broken and bruised are waged for reasons unknown to the beasts' … Man's petulance, his capacity for giving and receiving offence, combined with his extraordinary capacity for violence, makes his natural condition a very dangerous one, particularly when one takes into account the great divisions in human beliefs and ways of life. In short: 'Man, then, is an animal with an intense concern for his own preservation, needy by himself, incapable of protection without the help of his fellows, and very well fitted for the mutual provision of benefits. Equally, however, he is at the same time malicious, aggressive, easily provoked and as willing as he is able to inflict harm on others' … Man's life in the state of nature would thus indeed be miserable, unadorned, and short. It would not, however, be ungoverned by natural law or bereft of friendship as a primitive form of sociality. This is because man is indeed equipped by nature to know the natural law, even if he is not equipped to govern himself in accordance with it.

(Hunter 2001: 171–72, quoting Pufendorf; see also Hunter 2003; Hunter and Saunders 2003a, 2003b)

In other words, the anti-perfectionist approach to politics focuses almost entirely on the difficult but fundamental task of stopping humans from killing one another. Without this, the goal of an ever wider spread of civil peace would have remained an impossible dream. One cannot imagine how law or politics could seek to protect an endangered species of plant or animal or to control the effects of dangerous insecticides or to pursue any of the many tasks which today confront environmental law and politics if legal-political government had not been developed by using law and politics to bring to an end the spiral of violence that was civil war in early modern Europe. This is no easier than trying to imagine effective environmental law and politics in countries ravaged by civil war in the last twenty years, in Rwanda, say, or Bosnia, or, as we write this, Syria.

Sovereignty for legal-political government

Loughlin's (2003: 92) very basic definition of sovereignty follows closely from the points made in the last two sections: sovereignty is 'an institutional framework established for the purpose of maintaining and promoting peace, security, and the welfare of citizens'. On this matter too Loughlin regularly expresses his debt to Hobbes, though here he also expresses a debt to

Bodin's earlier arguments (for example, Loughlin 2003: 56–57, 79–80; see also Schmitt 2005).

Bodin's sixteenth-century interventions in France, particularly in the wake of the mass killings of St Bartholomew's Day in 1572 and the religious war which followed it (Saunders 1997: 73–76), were crucial to what Hobbes and his followers were able to do. Bodin was one of the first thinkers to use the term 'sovereignty'. Heinz Schilling argues (1986: 22) that for Bodin the term included 'modern forms of bureaucratic administration, a distinct territory within marked boundaries, and a concentration of political power in a supreme, central institution, usually the person of a king or prince'. For Bodin and the other French reformers with whom he worked (a group sometimes called the 'Politiques'), 'concentration of power is justified as the one acceptable alternative to religious civil war. Mutual butchery of citizens was Bodin's *summum malum*, the uttermost evil to be avoided at all costs' (Holmes 1988: 7).

Hobbes is justly famous for his determination to show that sovereign authority is not vested in a 'private man' or a group of 'private men' (1845b: Dedication). This is an early instance in *Leviathan* of Hobbes making the point that the sovereignty of the state resides in the office, not in the human individual or individuals who occupies or occupy the office. In this sense Hobbes thinks of sovereignty as a public common power which stops 'private men' – by keeping 'them all in awe' – from doing to one another 'a great deal of grief' in 'a war ... of every man, against every man' (1845b: 112–13). Hobbes is no less insistent that to become sovereign the individual (king, queen, prince, etc.) or assembly (parliament) who or which is most directly charged with running the office of the state must *publicly* be made sovereign by the covenants of the subjects (1845b: 161, 1845a: 101). This last point speaks directly to the idea that sovereignty is not an imposition on an unwilling population but is, in a more or less limited sense (the sense being limited by the level of security of the sovereign power, as we will discuss shortly), a bond between ruler and ruled, sometimes referred to as a 'contract'. Raoul van Caenegem (1988: 191) suggests that the operation of medieval law was a precursor of this aspect of sovereignty: 'The possibility for all free men and women ... of starting an action in the king's court, even against some powerful local personage, and of obtaining a judgement invested with all the authority of the crown meant a considerable curb on the lords and created a special and direct link between the people and the monarchy.'[5]

In this way, the sovereign state becomes a very particular type of 'person', the 'representative of all and every one of the multitude', the 'person' who or which always acts on behalf of the multitude (Hobbes 1845b: 171, 1845a: 140, 158). More than this, the sovereign state is sovereign because it is

---

5 Later in the same piece van Caenegem puts the point this way: 'the notion that the relation between rulers and citizens is based on a mutual contract ... means that governments have duties as well as rights and resistance to unlawful rulers who break that contract is legitimate' (1988: 210).

charged with achieving and maintaining 'the common peace and security' (Hobbes 1845b: 235). As Loughlin (2003: 59) puts it, 'the sovereign holds an office impressed with public responsibilities and for the realization of which he [or she or it] is vested with absolute sovereign authority'.

For the concerns of this book, it hardly needs adding, sovereignty is the authority that allows law and politics, in the person of the state, to govern the environment and climate change in a manner more effective than that which can be expected when there is no sovereign authority, when civil war has destroyed any chance of such authority. Sovereignty is the authority which, for example, was not available for governing the environment and climate change in places like Rwanda and Bosnia while civil war raged and is not, for the same reason, currently available for governing the environment and climate change in Syria (for more on the difficulties of governing when sovereignty has broken down, see Holmes 1994).

The state for legal-political government

We said above that the sovereign state is the dominant 'person' (in the widest sense of that term) in the legal-political system of government. This is the case because any serious attempt at rule must be strong enough to win and maintain a monopoly on the means of large-scale violence in the territory it seeks to rule if it is to attempt to move from being a 'serious attempt at rule' to becoming a sovereign state. This began to happen more often in the wake of the 1648 Treaty of Westphalia, which forced states, on pain of continued war, to recognize the legitimacy of every other state in their region, but even with this treaty in place each state's monopoly on the means of large-scale violence within its territory was often tenuous, especially in many fledgling states and/or in places where civil war or some other catastrophe led to the collapse of the state, led, that is, to what is sometimes called a failed state.

This is not to say that the notion of state did not come into existence until 1648. It is considerably older. Quentin Skinner (2002a: 369) proposes that it was during the 'revival of Roman law studies in twelfth-century Italy' that 'the word *status* came ... to designate legal standing of all states and conditions of men'. J.P. Canning (1988: 341) argues that, 'The thirteenth century marked the great turning-point in medieval political thought: an idea of the state was clearly acquired and located within an overtly political and this-worldly dimension'. He goes on to make the telling point that states as we now know them could not have emerged without an independent cohort of trained personnel (independent, that is, of the churches):

> [T]he process of political development was a gradual one ... What can be discerned is the emergence of politically organised communities (or peoples) with specific defined territories within which the internal and external sovereignty of rulers or governments was developed. Crucial to this process was the growth in the number of professionally trained

personnel required for the expansion of legal and judicial activity and of government ... The extent to which ecclesiastical jurisdiction retained a measure of autonomy in the late Middle Ages prevented the emergence of fully modern states.

(Canning 1988: 350)

In a similar vein, Skinner (2002a: 370) notes that the range of the term state expanded in the thirteenth and fourteenth centuries, such that, 'By the end of the fourteenth century, the term *status* was also in regular use to refer to the state and condition of a realm or commonwealth'. In the Italian city-states in the fifteenth and into the sixteenth century 'the common good', 'the cause of peace', and 'the happiness of the people' had to be achieved before a city could properly be called a state (Skinner 2002a: 372). In the later sixteenth and into the seventeenth century, as legal-political government gained ground in England, France, the Netherlands, and Germany,[6] the spread of 'practical political reasoning', particularly in 'advice-books for magistrates' and in the 'mirror-for-princes literature', '*status* and *stato* ... began to be used in new and significantly expanded ways' (Skinner 2002a: 374), finally giving 'the state' something like the meaning we have for it in this book.

This is a useful summary of important developments in the history of the notion of the state, but it does not tell us enough about what the state does in legal-political government. In short, the state is the vehicle by which the above two aims of politics are achieved – it musters and maintains the force needed to win and hold a monopoly of the large-scale means of violence within its territory and it provides the wherewithal by which the restrained type of politics (the administration, negotiation, diplomacy, etc.) keeps the territory's population from seeking recourse to violence against its own government. In this the state does what it has to do to meet Michael Oakeshott's very minimalist criterion for state activity: '"the activity of attending to the general arrangements of a set of people whom chance or choice ... brought together"' (Loughlin 2003: 79, quoting Oakeshott).

In saying what the state does we are obliged, we believe, to say something about how it does it. We will do so first by discussing Hobbes's account and then by discussing the account of a twenty-first century commentator on Hobbes (and on other early modern statist thinkers), Paul du Gay. Hobbes, as we saw in the previous chapter, usually calls the state the 'commonwealth' or refers to it by the Latin word '*civitas*'. For him, the state is the Leviathan, the focus of sovereignty, it is the absolute ruling force needed to stop humans from killing one another and to force them instead to work towards greater

---

6 Skinner (2002a: 373) says that 'The Erasmian humanists imported the same values and vocabulary [from the Italian city-states] into northern Europe in the early decades of the sixteenth-century'. He also highlights the sovereign state's normative insistence on civil peace: 'If the right *status* of a community is to be preserved, all factional advantage must be subordinated to the common good' (Skinner 2002a: 372).

civil peace. Crucially, the state's sovereignty includes the power of life and death over all subjects within its territory. Consistent with this power, one of the two main ways in which the state does its work is by instilling fear into its subjects. We will come to the second way after some more discussion of the role of fear.

While Hobbes says it might occasionally be 'dishonourable' (1845b: 79), excessive (1845b: 95), and ineffective (1845b: 285–86), without fear the state would achieve very little: 'Fear of oppression, disposeth a man to anticipate, or to seek aid by society: for there is no other way by which a man can secure his life and liberty' (1845b: 88). 'Fear and liberty', this is to say, 'are consistent'; 'generally all actions which men do in commonwealths, for *fear* of the law, are actions which the doers had *liberty* to omit' (1845b: 197, emphasis in original). This principle is as sound today as it was when Hobbes published *Leviathan* in 1651. Of course in most countries ruled by law and politics in the twenty-first century one would not expect citizens to fear their own state in the same way they did in the second half of the seventeenth century, when states sought to impose their strength on populations which had, for a long time, been killing one another on a grand scale over what those in law-and-politics countries today would see as minor differences of religious opinion or practice. This mass killings scenario is well beyond the experience of the vast majority of today's citizens in law-and-politics countries. But this is not to say that Hobbes is no longer relevant. Today the fear of the sovereign simply takes a different form: awareness of and respect for sovereign authority. As such, one most certainly would expect today's citizens to be aware that their state has the authority to punish wrongdoers, to impose order in difficult situations, and even to compel them to participate in military activities should the state come under attack by another state, and in this more limited sense today's citizens in law-and-politics countries are indeed controlled by fear.

In line with this more limited sense of fear of the state, one would expect these citizens to be aware that their state has the authority to establish a process whereby, for example, coastal erosion is monitored to determine whether housing or other development should be allowed in particular areas and to declare marine national parks to protect the natural heritage. One would expect them to be aware of their state's powers in these matters and to respect these powers even when they personally disagree with state agencies' decisions and declarations or when they stand to lose from them (possibly to lose their houses).

This 'different today but still similar' character of fear is equally true of the second of the ways the state does its work – instruction and persuasion. On the one hand, the list of six topics which Hobbes gives in Chapter 30 of *Leviathan* ('Of the Office of the Sovereign Representative') on how the state should instruct and persuade its subjects is unlikely to appeal to twenty-first-century citizens of states which have seen nothing for generations by way of existential threat:

[T]he people are to be taught, first, that they ought not be in love with any form of government they see in their neighbour nations, more than with their own … Secondly … that they ought not be led with admiration of the virtue of any of their fellow–subjects … nor of any assembly, except the sovereign assembly, so as to defer to them any obedience … appropriate to the sovereign only … Thirdly … they ought to be informed how great a fault it is, to speak evil of the sovereign representative … Fourthly … it is necessary that some such times be determined, wherein they [the subjects] may assemble together, and, after prayers and praises given to God, the sovereign and the sovereigns, hear those their duties told them, and the positive laws, such as generally concern them all, read and expounded, and be put in mind of the authority that maketh them laws … [Fifthly] … every sovereign ought to cause justice to be taught … Lastly, they are to be taught, that not only the unjust facts, but the designs and intentions to do them … are injustice.

(Hobbes 1845b: 336–40)[7]

But, on the other hand, when one takes into account the capacity, and the tendency, of twenty-first-century states to 'instruct' their citizens in, for example, the dangers of smoking and drinking alcohol, the need to lose weight and stay fit, the need to work harder, the need to save more for retirement, the need to sort garbage into different categories, the need to give up the use of plastic bags, and so on, one wonders whether the difference between what Hobbes was saying in 1651 by way of instruction and persuasion and what present-day states are saying is more a matter of different perceived dangers than of greater individual freedoms. At very least, we concede, today's items of instruction and persuasion are more subtle and in this way they serve as evidence for the proposition that the state has achieved quite a lot in the course of the last 360 years. The legal-political state as it operates now, protecting as it does an array of individual freedoms, allows the restrained type of politics to reach further in controlling its population than raw power politics ever could. As such, the state, as the embodiment of sovereignty and the focus of law and politics, is clearly the most important actor in the government of the environment and climate change today.

This brings us to du Gay's much more recent arguments (du Gay 2012). He worries that even when it is defined 'minimally' – 'as the political apparatus that delivers the governmental capacity needed to protect the members of a territorial population from each other and from external enemies' (du Gay 2012: 399) – too many people misunderstand the nature and purpose of

---

7 The preamble to the list is no less heavy-handed: '*the safety of the people*', in involving more than 'bare preservation', must entail 'a general providence contained in public instruction, both of doctrine, and example; and in the making and executing of good laws, to which individual persons may apply their own cases' (Hobbes 1845b: 322–23, emphasis in original).

the state. In 'the last three to four decades the idea of the state and the ideals of state service have been subject to extensive and near constant political, ideological, and theoretical criticism'. His most basic defence against such criticisms is a simple one: 'despite the claims of politicians, management consultants, human rights activists, and sociologists (to name just some) that the state as a bundle of institutions, purposes and conducts is, *inter alia*, an anachronism in a globalised world, an ideological disappointment, and a totalitarian threat to individual liberties and freedoms, it still remains difficult to imagine doing without it' (du Gay 2012: 399). He knows that such critics (like those we earlier saw Saunders challenge) are so enamoured of their own perfectionist moral codes that they fail to see that the state has its own particular morality, to do with maintaining a widespread appreciable civil peace.

In discussing the way in which these critics in fact rely on 'the norms of contemporary ethical and political culture', du Gay refers only obliquely to the techniques the state uses in doing its work. To 'put it mildly', he says, the critics are 'deeply suspicious of, or stand in outright opposition to, many of the key norms and techniques of conduct informing the activities of the state (authority, command, indifference, detachment, impersonalism, and so forth)' (du Gay 2012: 399). He renders less oblique this description of techniques – 'authority, command, indifference, detachment, impersonalism, and so forth' – by combining some historical points about the state's early modern development with some of Weber's early twentieth-century observations:

> For instance, the distinction between an office and the person holding that office became sharper and began to harden as a separate, highly structured domain of offices arose, and associated with those offices a greatly accumulated set of powers developed – resources, and instruments which were (and remain) not really under the effective personal control of those who happened to occupy the offices at any given moment.
>
> (du Gay 2012: 400)

This is to say that the distinction between an office, as a prescribed set of duties, and the person or persons who happen to hold the office – the person or persons charged with performing those duties – is one of the ways the state exercises its authority in a carefully indifferent, detached, impersonal manner, as we suggested earlier. In this, as with its other techniques, the state is working towards its distinctive moral goal, seeking an appreciable civil peace *across the entire population*, by which it is of course simultaneously seeking to minimize the danger of setting one part of the population against another by acting on behalf of one faction's interests against those of other factions. The state, du Gay points out, is thus able to exercise its authority over the population without having to continually remake its power of command for each different context:

[T]he authority of the state is both binding and content-independent: it forms a premise for the subject's action without that subject considering the merits of what it requires. The authority of the state cannot be seen as the authority of the people who constitute the subjects of the state, either individually or collectively. Therefore the state cannot be seen as the power of its citizens under another guise, as republican authors argue. Rather, the emergence of the state indicates a situation in which an alternative view predominates; one where the ends of civil or political association make it indispensable to establish a single and supreme sovereign authority whose power remains distinct not merely from the people over whom it is exercised, but also from whichever office-holders may be said to have the right to wield its power at any particular time.

(du Gay 2012: 400)

The state, this is to emphasize, is never to be confused with its subjects, it is not owned by them or controlled by them, even by those who occupy the offices which operate state power. In Blandine Kriegel's words, 'public offices belong neither to lords nor to a prince, nor to a state, for they are the state itself' (1995: 26–27). To put this crucial point in yet another way, the state is separate from 'the will of the people', and it is separate from 'the personal and moral characteristics' of the human individuals 'who exercise state power' (du Gay 2012: 401). This is because the 'exercising' of state power is actually done by the office, not by the office holder. The morality of the office is confined entirely to the promotion of as wide an appreciable spread of civil peace as possible; it may or may not coincide with the personal moral convictions of the individual occupying it at any given time.

An individual who works for a state environmental agency, for example, might well be a committed environmentalist in his or her private life. But when occupying the state office the individual ceases to be an individual with private moral convictions about the environment. In this capacity – that is, when she or he is on duty – she or he must act only in accordance with the morality of the office, even if a particular decision on the environment – taken in the light of a formal judgment about the best strategy in the circumstances for maintaining or enhancing the widest possible appreciable civil peace – is not consistent with her or his private beliefs about the good of the environment.

This aspect of the state is perhaps easier to grasp if one continually bears in mind the fact that the state can never stray far from the most fundamental of its tasks: security. The 'default setting' of the state (Hunter 2005), as it were, is the provision of security against threats to its population, particularly against threats posed by civil violence or by other states, but also against threats posed by tsunamis, floods, earthquakes, major fires, major droughts, and even financial crises. Drawing on Hunter (1998), du Gay puts the matter like this:

Under conditions of peace the sovereign state, in its guise as security state, can almost seem to disappear, and the state becomes the addressee of a wide range of additional demands and expectations. As soon as the security envelope is threatened, however, whether internally (incitement to insurrection, domestic terrorism, economic crisis) or externally (foreign terrorism, invasion, economic crisis) then civil liberties and rights are retracted to the extent that is necessary to protect the space within which they were unfolded in the first place ... It should therefore come as no surprise – in fact it should be taken as evidence of its 'stateness' – that contemporary liberal democratic states default to their 'foundational' security setting when under threat.

(du Gay 2012: 402–03)

Before we discuss the economy, we think it wise to say a little more about why the state needs total authority. In times of peace the state appears to be but one actor among many on the governmental stage, all of whom might be thought to deserve at least some input into major decisions about a country's direction. By this argument, political parties, independent groups of activists, churches, or corporations, among others, might be seen to be vital sources of authority on many issues.

But as du Gay points out, drawing on the work of John Dunn, 'the state cannot possibly undertake its core functions of pacification and security unless it can decide for itself, "without internal impediment", what can be publicly expressed, or just who can own what, and why' (2012: 403, quoting Dunn). More than this, the state must have this supreme power *on behalf of all other actors*. Again du Gay employs Dunn to emphasize that only the state is in a position to

'judge the degree of jeopardy in every instance. The state carries, and must carry, the authority of its own subjects' will and choice to make that judgement on their behalf and to act, decisively, upon it. Indeed, each subject has a right against every other that it [the state] should do just this'.

(du Gay 2012: 403, quoting Dunn)

For those still unconvinced about the necessity of supreme state authority, du Gay borrows a more pointedly normative element from Raymond Geuss:

'[T]here is *always* going to be a gap between the political power of the state and the effective powers of the populace, and, on this argument, that is a good thing ... in a world populated by *other* states, many of them predatory, it is essential for the minimal self-defence of a certain population that it be organized as a state, or one might think that it was necessary [as recent events indicate all too clearly] to have an independent power that could intervene in the economy to prevent it from self-destruction.

(du Gay 2012: 403, quoting Geuss, emphasis in original)

The economy for legal-political government

In the introduction to this chapter we used the term 'the economy (as a separate domain)'. The parenthetic addition helps to make clear that while the creation, maintenance, and use of wealth is as old as government itself (as indeed are many of the techniques associated with the use of wealth, including mundane taxation as well as more brutal plundering by invading forces), the notion of a distinct sphere of knowledge concerned entirely with the creation, maintenance, and use of wealth towards the goal of an ever wider appreciable spread of civil peace (particularly its prosperity component) is in fact relatively recent. What is most important to us here is that as this distinct sphere of knowledge emerged, in the eighteenth and especially the nineteenth century, it quickly took as its object 'the economy'. As soon as this sphere of knowledge emerged it became a domain in which legal-political government sought to dominate, to the point where 'the economy' became a separate domain of legal-political governmental activity, something which law-and-politics governments had to learn to control if they were to govern effectively. This is to say that the economy became not just a concern of the state but one of its principal foci.

Over the course of the late seventeenth and into the eighteenth century the state eagerly tried to build existing areas of knowledge – particularly a body of knowledge known as 'counsel on trade'; in 1729 William Defoe noted that '"if any one nation could govern Trade, that nation would govern the World"' (Walter 2011: 50–51, quoting Defoe) – into what we might now call a strategy for governing the economy. Using law and politics, the state, as the sovereign power, sought the type of economic knowledge by which it could continually refine its attempt to achieve an ever wider appreciable spread of civil peace among its subjects. It was the nineteenth-century work of David Ricardo, Walter argues (2011: 91–111; see also 2008a, 2008b), that allowed the state to formally recognize the economy by name. Painfully, through collapses like the famous South Sea Bubble of 1720 (Walter 2011: 15), the state also had to learn to deal with the other side of the process of wealth creation, the side by which its failure to control the economy could lead to less peace and often far less prosperity. Of course it is still doing just this, with more and less success, as the recent global financial crisis attests (Wickham and Bryan 2012). Taken on the whole, however, the state can be said to have achieved much greater command over the economy than it once had, though of course still nothing like as much command as it would like. The state, this is to say, still has no choice but to practise a remarkable level of flexibility, making calculations on a regular basis (daily or even hourly in times of crisis) about which levers to pull, which decisions to leave entirely to the market, etc.

Because of the economy's importance, the state cannot suddenly stop worrying about it when it is dealing with the government of the environment and climate change. This is why we suggested earlier that in governing the environment and climate change the state must deal with an ever shift-ing balance between the benefit of humans (towards as wide an appreciable spread of civil peace as possible) and the maintenance of nature. Were it to forget the prosperity component, as it dealt with any environmental issue, it could contribute to instability and hence threaten peace. Equally, were it to let economic considerations completely dominate environmental ones it could also contribute to instability and threaten peace, whether because of natural disasters (including environmental degradation), or the opposition of environmentally concerned citizens, or both.

Interests for legal-political government

As legal-political government gained strength, at the end of the seventeenth century and into the eighteenth, its new way of doing and understanding law and politics allowed its principal actor, the sovereign state, to imbue more and more subjects with the confidence to treat religion as a private matter and to understand public morality as the widest possible appreciable spread of civil peace. With this shift came a new way of thinking about how vari-ous actors might position themselves in relation to other actors. In the first instance this applied to the state itself, in its dealings with other states and its dealings with other large-scale actors, such as the church and emerging com-mercial enterprises. As Walter (2011: 21) puts it, 'By the early seventeenth century, the terms "reason of state" and "interest" enjoyed wide currency in western Europe, not least because they opened up matters of state to public scrutiny'. The notion of interests, he goes on, 'was a species of political knowledge' that helped to make 'the state and state policy both visible and calculable' (Walter 2011: 24).

While the notion of interests is our main concern here, we should say some-thing about reason of state, partly because it is, as Walter says, a companion of the notion of interests in the development of legal-political government, and partly because it allows us to acknowledge a debt to Michel Foucault's work on the history of government, work built around the notion of governmentality.[8] On 'reason of state' in particular, Foucault says:

> To put it very schematically: the art of government finds at the end
> of the 16th century and the beginning of the 17th its first form of

---

8 In his work on governmentality, Foucault (see esp. 1979) argues that from about the sixteenth century a new way of thinking about the state emerged which allowed it to be seen as an inde-pendent entity, with its own conditions of operation. In turn this led to the emergence of a spe-cific object of state government, 'the population', which led to the birth of specialist knowledge endeavours, like political economy and statistics, which themselves encouraged thinking about new forms of liberalism. For more discussion of what Foucault was trying to do with the notion of governmentality, as an aid to the analysis of government, see esp.: Barry et al. (1996); Burchell et al. (1991); Dean (1999); Dean and Hindess (1998); Miller and Rose (1990); Osborne (1998); Rose (1989), (1993), (1996); Rose and Miller (1992); Walters (2012).

crystallisation: it organises itself around the theme of the reason of State, understood not in the negative and pejorative sense we attribute to it today (namely as that which infringes on the principles of law, of equality and humanity in the sole interests of the State), but in a full and positive sense: the State is governed according to rational principles which are intrinsic to it and which cannot be derived solely from natural or divine laws or the principles of wisdom and prudence; the State, like Nature, has its own proper rationality, even if this is of a different sort. Conversely, the art of government instead of seeking its foundation in transcendental rules, cosmological models or philosophical-moral ideals, must find the principles of its rationality in that which constitutes the specific reality of the State.

(Foucault 1979: 14)

Returning to the notion of interests, Albert Hirschman, writing before Foucault published any of his 'governmentality' material, also chose the sixteenth century as the starting point for an investigation. He uncovers a very early form of 'interests' circulating as part of attempts to quell the passions of civil war. It was proposed at that time that it would be less dangerous if humans paid more attention to their 'interests' instead of giving 'free rein to their passions'. As such, the notion was born broad:

When the term 'interest' in the sense of concerns, aspirations, and advantage gained currency in Western Europe during the late sixteenth century, its meaning was by no means limited to the material aspects of a person's welfare; rather, it comprised the totality of human aspirations, but denoted an element of reflection and calculation with respect to the manner in which aspirations were to be pursued.

(Hirschman 1977: 32)

In this way, Hirschman (1977: 43–44) continues, in the sixteenth and seventeenth centuries 'interests' were considered by some to be a 'hybrid form of human action … exempt from both the destructiveness of passion and the ineffectuality of reason'.

Like Walter, Hirschman (1977: 40) shows that later in the seventeenth and into the eighteenth century, as the economy gained strength as a separate domain, the notion of interests also took on a decidedly narrow, economic meaning: 'the beginnings of economic growth made the "augmentation of fortune" a real possibility for an increasing number of people' which led to a 'narrowing of the term "interests"'. This 'narrowing', it should be noted, was not seen as a negative; from this time onwards 'the term "interests" actually carried – and therefore bestowed on money-making – a *positive* and *curative* connotation deriving from its recent close association with the idea of a more enlightened way of conducting human affairs, private as well as public' (Hirschman 1977: 42, emphases in original).

The notion of interests that became important to legal-political government, and which remains important today, lies between the broad and the narrow meanings. As Hirschman (1977: 48) puts it, because it was predictable and constant, the pursuit of interests as a 'dominant motive for human behavior' was seen to provide a 'realistic basis for a viable social order'.[9] Hirschman (1977: 51) goes on to propose that the 'benefits' of predictability 'loomed largest when the concept was used in connection with the economic activities of individuals'. In adopting and adapting both meanings of the notion, the prudent state thus sought (and still seeks) both to foster greater prosperity for its population and to keep a close eye on those 'interests' within its population, and/or those within other states, that might foster civil unrest more than they foster greater prosperity.

In another passage, Hirschman (1977: 51) allows the notion to expand such that it becomes 'the public interest': 'After the Restoration and during the debate on religious toleration, there was much discussion about the advantages that might accrue to the public interest from the presence of a variety of interests and from a certain tension between them.' In Chapter 5 we will show how the notion of 'the public interest' has become important in the legal-political government of the environment and climate change. For the time being we will use just a brief discussion of this aspect of the operation of law to expand upon some of the points about interests made above.

The notion of public interest in law, especially as it deals with the environment, has become especially important as a rhetorical device. Doreen McBarnet (1981: 161), for example, suggests that as a rhetorical device the public interest is essential to the survival of the common law, particularly at the level of court proceedings. The notion of the public interest, this is to say, affords lawyers rhetorical opportunities to persuade a court to act on an established legal position *and* it allows the court to be so persuaded. In other words, as a rhetorical device the notion of interest lets the common law operate in a more sophisticated manner than the traditional liberal picture of it would allow. Instead of reducing the public interest to the interests of individuals, in line with some supposed commitment to classical liberal modes of calculation, the common law is calculating pragmatically, as it has long done.[10] Courts have careful procedures for taking the public interest into

---

9 Hirschman quotes from a seventeenth-century pamphlet by Marchamont Nedham, pointedly titled 'Interest Will Not Lie', to help make his point: '"If you can apprehend wherein a man's interest to any particular game on foot doth consist you may surely know, if the man be prudent, whereabout to have him, that is, how to judge of his design"' (Hirschman 1977: 49, quoting Nedham). The entry for Nedham in the *Oxford Dictionary of National Biography*, by Joad Raymond (2008), recounts his skill as a pamphleteer and as a Machiavellian thinker, which was such that he was able to appeal, at different times, to both sides of the English Civil War, though was fondly regarded by neither.

10 In praising the seventeenth-century English jurist Matthew Hale, David Saunders (1997: 44) says that by Hale's way of thinking, 'common law reason offers no ... magic instant. On the contrary, it is the slow and unplanned aggregate of many judgments.'

account, including recognition of specialist 'public interest' advocates in cases concerning the environment (more on this in Chapter 5).

This is clearly an example of law operating as part of legal-political government, a part with a particular procedure for recognizing and fostering interests. This 'recognizing and fostering interests' is part of a broader strategy, whereby governments in law-and-politics countries put considerable effort into such recognizing and fostering through the development of particular policies and programmes. In this way, the recognizing and fostering of interests produces yet more interests. But of course, as Barry Hindess points out (1986: 17), this process is always contestable and therefore contingent upon a range of factors that go beyond the motivation or claims of the individuals and institutions who and which are asserting their interests. Furthermore, 'Interests are the product of assessment. They do not appear arbitrarily, out of nowhere, they are not structurally determined and they cannot be regarded as fixed or given properties of actors' (Hindess 1986: 25).

The nature of the relation between law and politics for legal-political government

The relation between law and politics in legal-political government is nearly always a tense one, sometimes very tense. It was extremely tense in the early modern period, when legal-political government was in its infancy, which is hardly surprising in a climate of mass killings born of religious hatreds and other extreme passions, a matter to which we will return presently. While of course it is not as tense in twenty-first century law-and-politics countries, it is tense nonetheless. The tension, in other words, is not just about the difficult circumstances of the birth of legal-political government, it is also productive tension, that is, it is a tension which helps to make legal-political government work.

For example, if a group of cliff-top residents were to be told that their houses were under grave threat of inundation by rising seas, probably caused by climate change, and therefore must be demolished, the issue would be fought out in the courts *and* it would be fought out by direct political means (protests to the local council and/or to higher authorities) and by indirect political means (media bombardment, etc.). In rare cases it might also spill over into the darker side of politics (as in the famous Erin Brockovich case). Those engaging in the political activities, especially the indirect and/or darker ones, would likely find the legal effort slow and cumbersome, especially if some of the threatened demolitions take place. Those taking responsibility for the legal activities, on the other hand, would likely find the political activities, especially the darker ones, a hindrance, possibly worse. But almost certainly the issue would be resolved, one way or the other (even if only temporarily), in the spaces carved out by the tense interactions between law and politics. It has always been this way for legal-political government.

Hunter (2001: 83–84) talks of the early modern union of law and politics as the uneasy coming together of two 'independent strategies … each draw-ing on the intellectual resources at its disposal in order to forge instruments capable of meeting the challenge to governance posed by religious civil war'. Each of these 'strategies' viewed itself, at that time, as it still does now, as the dominant partner in the union, and with good cause. On the one hand, as we saw in an earlier section, law, the constituting law in particular, has a claim that because it guarantees the state's sovereignty – the supreme power it needs to operate – it therefore dominates politics. Yet, as we also saw above, politics, because of its raw energy, its capacity for deadly violence, has a claim that because it is the engine of all forms of rule, it dominates law.[11] We think it best to sit on the fence. While politics may be technically in a position to assert primacy over law, because its capacity for deadly violence is so great as to be unstoppable, law's restraining influence is vital. Without law's influence politics could destroy even itself, by destroying all participants. In the end, even as they snarl at one another, we think it best to conclude that law needs politics for its raw energy and politics needs law for its remarkable capacity for restraint (see especially Wickham and Bryan 2012: 197–98).

Having described the relation between law and politics in the legal-political system in terms very similar to those we used in an earlier section – where we argued that in legal-political government the law must be strong enough to temper politics in such a way that they work together, however easily or uneasily, such that in any country in which law is an equal or near-equal force, the great majority of the population of that country will have the opportunity to appreciate and exercise the freedoms that the genuine main-tenance of peace, security, well-being, and prosperity will deliver to them (even if they choose not to exercise those freedoms) – it is now possible for us to add a further explanation onto an important point we introduced in that section, about the differences between legal-political governments and those which may have a commitment to peace, security, well-being, and prosperity but do not have an *appreciable* commitment. We made clear on that occasion that in countries ruled by governments which rely much more heavily on politics than they do on law, the great majority of their populations will not be in a position to appreciate and exercise the type of individual freedoms we discussed earlier (with considerable help from Kriegel).

Our further explanation is that the great majority of the populations of these countries will not be in a position to appreciate and exercise the individ-ual freedoms because the dominance of politics in these countries means that

11 Loughlin (2003: 91–92) puts the strong claim for the primacy of politics over law in this way: '[W]e should not forget that law's function – the duty of all officers of the law – is to maintain and bolster the sovereignty of the state. The constitution, as Justice Robert Jackman once explained, is not a suicide pact … Law plays a critical role in explicating in the form of rules, regulations, rights, and responsibilities the character of sovereign authority. But if public law is to be taken seriously, we need to recognise that, not withstanding certain rhetorical flourishes about the appeal to "higher", "fundamental" or even "natural" law, the determination of the limits to sovereign authority, even when articulated by courts, must be political.'

its raw power (that is, the raw power of politics, possibly made even stronger by ideological or religious allegiances) will be so great as to make individual freedoms impossible to guarantee. These freedoms may well be widely desired and some of the governments in these countries may even, albeit only occasionally, make an attempt to introduce at least some of them, but the people being ruled in this way will know from experience that it would be impolitic of them, at very least, to seek to explore these freedoms too far, or, at worst, would put their lives in peril. In this way, the subjects of this type of rule quickly work out that their governments' power is a negative *power over* more than it is a positive *power to*. This explanation allows us to make even clearer the distinction we are drawing here. On the one hand, in law-and-politics countries the 'appreciable' aspect of legal-political governments' commitment to peace, security, well-being, and prosperity, in convincing the people in their countries that they are free to participate in political issues, produces a very complex political landscape, with many different interests likely to be active in most political issues. On the other hand, in countries where governments have a very different type of commitment to the four goods of civil peace (peace, security, well-being, and prosperity), in the manner just described, the political landscape will be far less complex, with a much more predictable set of interests likely to be active in most political issues.

Before we move on we must stress that none of this renders such governments' commitment to peace, security, well-being, and prosperity meaningless; far from it. In countries where politics dominates in the manner described above – whether in China, where raw politics is made stronger by ideological allegiances, or in one of the theocratic countries of the Middle East, where raw politics is made stronger by religious allegiances – governments are genuinely committed to peace, security, well-being, and prosperity and participate in international arrangements on this basis. They may well seek peace, security, well-being, and prosperity for reasons that are in some way foreign to those who live under legal-political rule, but this does not make their commitment to these goods less genuine.

Our final point on the relation between law and politics is that the anti-perfectionist thinking behind legal-political government does not rely on the idea of a third force, a force which can trump both law and politics. As Hunter (2001: 66–67) puts it, 'early modern politics and jurisprudence cannot be explained either in terms of a theory of ... society or in terms of a philosophy of subjectivity'. In this way, Hunter is aiming his guns at those early modern metaphysical perfectionists who, still inspired by Aquinas's reworking of Aristotle and Plato (as many scholars still are today), were determined to undermine the then new form of government by re-establishing the supremacy of a perfecting force to which humans and their laws and politics must be subjugated, usually in the name of nature, the universe, pure being, and/ or God. As Hunter argues, the 'pursuit of pure rational being ... drove metaphysical philosophy for Liebniz through Wolff to Kant and beyond' (2001: x). Employing 'pure concepts of morality and reason', the perfectionist rearguard was determined, Hunter argues, to reverse the 'desacralisation' of law and politics which had already been achieved by the anti-perfectionist thinkers (see, for example, Hunter: 2001: 74, n.34). This rearguard claimed for itself the 'authority to limit the governance of the earthly city in accordance with the laws of its divine archetype, thereby advancing the interests of the academic-clerical estate' (Hunter: 2001: 28). Hunter highlights the efforts of the two German anti-perfectionists mentioned several times, Pufendorf and Thomasius, to keep the metaphysicians at bay. For example, here is what Thomasius had to say on the matter:

> '[N]othing has been more responsible for derailing man's natural pursuit of a long and happy earthly life than the mixing and confusion of … two kinds of truth; from this have arisen shameful exercises of priest-craft with all their attendant misery of religious tyranny and conflict … [especially] those who inherited the pagan philosophical conception of nature and mixed it with the Christian doctrine of creation – that is, the metaphysicians. Ensnared by "Platonic fables", the metaphysicians not only produced a bastard philosophical-theological conception of a creation divided into visible and invisible things, they also used their alleged insight into transcendent being as the basis for doctrine-mongering and religious oppression.'
>
> (Hunter 2001: 85, quoting Thomasius)

In our terms, Thomasius was working towards a new form of anti-perfectionist knowledge, a form 'suited to the jurists and politici of the desacralised state' (Hunter 2001: 10 n34), that is, a form which would support legal-political government by enhancing 'the new doctrines of territorial sovereignty and desacralised politics in a specific intellectual deportment' (Hunter 2001: 28).

Before concluding the chapter, we wish to stress that the early anti-perfectionist thinkers were not, in combining law and politics in the way we have described, working from a blue-print. The circumstances of ongoing civil war meant that they had to be extremely flexible: '[O]nce it became abundantly clear that the religious wars were incapable of theological adjudication or military-political termination [the early modern] jurists developed a series of measures designed to end the conflicts by securing the coexistence of the confessions' (Hunter 2001: 82, n.34). These moves 'were not the result of transcendent philosophical reflection whose culmination would come in the democratic natural law theories of the Aufklärung … they arose as unplanned consequences of a whole series of juridical improvisations undertaken by anonymous political jurists seeking the legal-political bases of social peace' (Hunter 2001: 84). This was the temper of the times, and, in terms of the relation between law and politics, it still is.

Conclusion

The chapter has detailed the role of each of law, politics, sovereignty, the state, the economy, and interests in building and maintaining legal-political government as a single entity, an entity which leans towards anti-perfectionist thinking but which is never immune from the lure of perfectionism. So far, however, we have said very little about other elements which are drafted into legal-political government. Some are older elements which perhaps now seem peripheral to governmental concerns, such as aesthetics, while some are newer and usually seem vital to these concerns, such as science. These two 'other elements' will come into their own in Chapter 3, but we need to introduce them here before offering our closing remarks.

Aesthetics will be seen in the following chapter to play an important role in the government of the environment, in both its 'higher' or elite form – in which the environment, sometimes simply as 'nature', is the subject of literature and painting – and its 'lower' or more populist form – in which the environment, whether as nature per se or as the site of sporting contests or recreational pursuits, is made available in ways in which it previously was not, for strolling, hiking, swimming, mountaineering, recreational hunting and fishing, surfing, golf, yacht racing, picnicking, cycling, driving, riding, and so on (this list could be made very long in law-and-politics countries in the early twenty-first century).

Of course the reader will already know that science plays an important role in the government of the environment and climate change. Since the rise of science as a concern of governing elites in early modern Europe (the formation of the Royal Society in 1660 is sometimes used to mark a beginning point), echoing a feature of ancient Greece and Athens and other ancient civilizations, it has come to be regarded by all governments in law-and-politics countries as a vital source of knowledge. In this way, science is seen not only as an aid to understanding and controlling the natural environment in a general manner, but mainly as an aid to understanding the particulars of economies, wars, industries, education, leisure, health, food, government itself, and, of course climate and the way it changes.

The narrative which drives the book has now been put in place. This chapter has embedded the book's key question – how do law and politics, entwined in various complex relations with sovereignty, the state, morality, religion, economy, aesthetics, and science, attempt to govern the environment and climate change? – more firmly in its historical context. In the harsh conditions generated by a plethora of civil wars in sixteenth- and seventeenth-century Europe, a new mode of government emerged – legal-political government – partly as the result of deliberate strategies and partly as the result of various contingencies. This new mode was based on, and contributed to, a new strain of anti-perfectionist thinking about humans, that is, thinking which rejected the necessity of trying to perfect humans, something which was deeply entrenched at the time, whereby rival religious doctrines, each with its own vision of good government, were determined to force their opponents either to submit to their particular form of perfection or, just as often, to kill them.

In these circumstances, legal-political government did not, of course, spring forth fully formed. It developed slowly between that era and our own, with many aspects of older modes of governing proving harder to shake off than others. Many such aspects are still powerful today. Among the more noticeable of the aspects which legal-political government has had to continually try to shrug off, with varying degrees of success in different places and at different times, is the idea that government must comply, or seek to comply, with timeless, universal moral codes, whether of religious or other heritage. In this way, legal-political government has worked hard to replace the stranglehold of the idea that all government must be driven by an eternal conscience, sourced in an external, supposedly timeless and universal morality, a conscience, that is, external to law and politics. While this insistence on the centrality of an externally sourced eternal conscience continues to linger, legal-political government mostly rebuffs its entreaties and dismisses its central argument – that government without a conscience is doomed to amorality and/or immorality. Far from retreating into amorality or immorality, legal-political government uses an alternative normative compass, by which it has introduced and entrenched its own, this-worldly and public, form of conscience. This is the conscience of civil peace, a conscience which compels law and politics to drive government to satisfy one goal above all others: the ever wider appreciable spread of civil peace.

This is the inescapable condition of legal-political government's ongoing interaction with the environment and climate change. In seeking to govern the environment and climate change, legal-political government must tread a careful path, one whereby it can pursue a wider spread of appreciable civil peace at the same time as, indeed as part of, enhancing the natural environment, of which humans are part. In treading this careful path, the legal-political government of the environment and climate change should never lose sight of the brute facts we stressed at the start of the chapter: this form of government is not government by religion-and-morality, it is not government by private moral conscience, and it is not government by lofty ideals; nor is it perfect, it has a long record of serious failures, and its capacity to deliver on its norm of appreciable civil peace is not limitless.

3 The environment (slowly) emerges as an object of legal-political government

Introduction: when was the name 'the environment' first used to mean a thing which governments should protect?

While legal-political government itself emerged in sixteenth- and seventeenth-century Europe, the environment did not become a central concern of this or any other form of government anywhere in the world until the second half of the twentieth century. This is primarily because neither the entity 'the environment' nor the associated notion of 'ecology' came into clear focus *as objects of legal and political concern* until the second half of the twentieth century.

In the last chapter we stressed that this fact does not mean that *what we now call the environment* was ungoverned for thousands of years. We pointed out then that if the words 'govern' and 'environment' are used in their broad, loose senses, briefly leaving aside the narrower, more exact meanings at the centre of our book's concerns (meanings to be detailed very soon), one can sensibly talk about the governing of the environment in any period of history. But this does not clear up the matter of when the stricter present-day sense of the term 'environment' first came to prominence.

We can clear up this matter in two steps: first, by using the *Oxford English Dictionary* (*OED* online) to specify the narrower, more exact meanings of the terms 'environment' and 'ecology' (the meanings we use in this book), which will also allow us to specify when these more exact meanings first came into use; and second, by discussing an example of 'governing what we would now call the environment' from the thirteenth century, to show what can happen when 'environment' and its cognate terms are used without proper regard for their historical context.

While the *OED* notes the use in England in the first half of the twelfth century (by Anglo-Norman speakers) of the root Middle French word *environnement*, in the sense of 'proximity', and notes the use of this word in a different sense (the 'action of surrounding something' (*OED* online)) in 1487, it lists no usage of the English word 'environment' before 1603, when it, like its Middle French counterpart, was used to mean the 'action of surrounding something' or 'the state of being encompassed or surrounded' (*OED* online), a definition which remained active into the twentieth century. But this of course is still far removed from the meaning of 'environment' as it is used in our term 'the legal-political government of the environment and climate change'. For that meaning we have to look much closer to our own era.

Perhaps one might glimpse something of this meaning in the definition 'physical surroundings or conditions in which a person or other organism lives … or in which a thing exists' (*OED* online), the earliest instance of which the *OED* puts at 1855 (and which is still current), but even this definition is not right. In fact, the *OED* takes us back only to 1948 for the earliest instance of the more relevant of the two definitions of 'environment' we consider central to our book (both of which, of course, are still very active today) – the 'natural world or physical surroundings in general, either as a whole or within a geographical area, esp. as affected by human activity' (*OED* online). The other (only slightly less) relevant definition of the pair, the earliest example of which the *OED* traces to 1936, is more pointedly political: 'The social, political, or cultural circumstances in which a person lives, esp. with respect to their effect on behaviour, attitudes, etc.' (*OED* online). Complementing this couplet of definitions are terms in which 'environment' serves as a qualifying word, such as 'environment editor' (earliest instance 1970), 'environment ministry' (1970), 'environment movement' (1970), 'environment policy' (1968), 'environment group' (1960) (as in 'an organization devoted to the study or protection of the natural environment'), and the important adjectives 'environment-conscious' (1935), defined as 'sensitive to the state of the surrounding environment', and 'environment-friendly' (1982), defined as 'sensitive to and in harmony with aspects of the (external or human) environment' (*OED* online).

As for 'ecology', the *OED* offers three definitions which might be helpful to us. First, ecology is the 'branch of biology that deals with the relationships between living organisms and their environment'. The dictionary lists 1876 as the earliest usage of this definition, though it notes that it was at that time still only available as a technical term, as marked by the use of the 'œ' diphthong ('The great series of phenomena of comparative anatomy and ontogeny, palæontology and taxonomy, chorology and œcology' (*OED* online)). For a usage of this definition closer to that which we regard as its common usage in today's debates about environmental government, one has to wait until 1904, by which time its spelling had dispensed with the diphthong: 'The study of plants that grow together, forming plant associations, in some respects the most interesting part of Ecology' (*OED* online).

The second definition, which the *OED* says is 'chiefly sociological' is: 'The study of the relationships between people, social groups, and their environment; (also) the system of such relationships in an area of human settlement. Freq. with modifying word, as cultural ecology, social ecology, urban ecology' (*OED* online). The earliest usage of this definition was in 1908, in the *American Journal of Sociology*: 'Human ecology, a study of the geographic conditions of human culture.'

In terms of our book, the third definition is certainly the sharpest: 'The study of or concern for the effect of human activity on the environment; advocacy of restrictions on industrial and agricultural development as a political movement; (also) a political movement dedicated to this' (*OED* online). Significantly, the earliest usage of this definition is listed as 1970, when the *Nevada State Journal* ran the following headline: 'It's the top topic for everyone now – ecology.' The *OED* also notes that compound terms relevant to our project, like 'ecology movement' and 'ecology symbol', were not used before 1969.

Having established some definitional solid ground, we move now to our thirteenth-century example of 'governing what we now call the environment'. Chris Besant notes that when the *Magna Carta* was 're-issued by Henry III in 1217, its schizophrenic character was made manifest as the Charter was promulgated in two documents: a *Charter of Liberties* and a *Charter of the Forest*' (Besant 1991: 160). In this document 'forest' was defined as land not subject to the tenurial system and so not subject to the common law, 'an area designated by the Regent to be within his special protection' (Besant 1991: 161).

For Besant, the *Charter of the Forest* is a de facto piece of environmental legislation: 'While on its face forest law is a system of game laws, in substance it imposed dramatic and effective restraints on the exploitation of natural resources ... [it] defined an equilibrium between human resource exploitation (husbandry, agriculture, settlement) and a natural resource ecosystem' (Besant 1991: 161). Furthermore, Besant argues, the protection of the 'natural resource ecosystem' remained the main goal of forest law, as John Manwood's 1598 *Treatise and Discourse of the Laws of the Forest* demonstrates:

> 'And those woods of covet that are felled by license, yet the springs thereof must be carefully preserved, that they may grow to be covets again in a short time, for they have license to fell their woods, yet they had not license to destroy them.'
>
> (Besant 1991: 162, quoting Manwood)

We think Besant is looking at this example of thirteenth-century forest law wearing twentieth-century spectacles. We propose that the 'equilibrium' described as that between 'human resource exploitation (husbandry, agriculture, settlement)' and 'a natural resource ecosystem' is better thought of not, as Besant suggests, as a precursor of successful environmental law per se – the notion of a 'natural resource ecosystem', our *OED* research makes clear, was unknown in the thirteenth century – but rather as a very early step on the road towards the much more imperfect project we call the legal-political government of the environment and climate change, a project which, as we said earlier, must continually attempt to control the ever shifting balance between the benefit of humans (towards as wide an appreciable spread of civil peace as possible) and the maintenance of nature.

More than this, while protected forested area is undoubtedly *what we would now call the environment*, we do not think the *Charter of the Forest* of 1217 was concerned with 'the maintenance of the environment' in anything like its modern sense. Instead, we think it was wholly and solely concerned with the other purpose Besant notes (without seeing what we regard as its import): to provide to those trying to make a living from the land (the vast bulk of the population) 'relief from the oppressions of the forest law' (Besant 1991: 161–62).

We have now confirmed our point that the legal-political government of the environment cannot be said to have begun until, first, the most basic combination of elements of legal-political government – law, politics, sovereignty, and state, alongside morality and religion – was in place (by the eighteenth century), and, second, the environment/ecology had emerged as a distinct object of such government (not until the twentieth century). We can now press on with the story of how an unlikely intersection of aesthetics and science, formed around the notion of nature, eventually produced a particular object for legal-political government which (to use the formulation for the last time) we now call the environment.

We will tackle this task in four main sections. In the first we will show how a new understanding of nature emerged in the late eighteenth and the nineteenth century, partly in place of the sixteenth/seventeenth-century understandings we highlighted in our opening chapter and partly to supplement them. The other three sections, in discussing the role of the two elements introduced at the end of the previous chapter, aesthetics and science, will concentrate on the nineteenth and the early twentieth century. These two elements, intersecting as they did across this period with religion and morality (as, to varying degrees, they still do), were the main vehicles for the new understanding of nature and in this role they drove the emergence of the environment as an object of legal-political government, eventually delivering this object into the path of law, politics, state, and economy. In the second section we will deal with the importance of aesthetics to the rise of the new understanding of nature. In the third section we will focus on the considerable degree of overlap between aesthetics and science in the nineteenth century. In the fourth section we will deal with the importance of science to the rise of the new understanding of nature. In the conclusion to the chapter we will pose two points by way of linking it to those chapters that preceded it and those that will follow it: one point will be concerned with the way science overtook aesthetics in the first half of the twentieth century, just as the environment was emerging as a distinct object of legal-political government; the other will be concerned with the fact that in spite of this both aesthetics and science can still be said to play strong parts in the legal-political government of the environment and climate change.

The new understanding of nature which emerged in the late eighteenth and nineteenth century

We said quite a lot about the notion of nature in the introductory chapter, particularly in discussing the importance to our project of the history of natural law in the sixteenth and seventeenth centuries. In that discussion we talked about the emergence of two rival versions of nature, particularly as they were expressed in two rival accounts of human nature. On one side of this rivalry was a perfectionist understanding of nature, by which either (or both) a perfect God or a perfect Nature produces humans who will always strive towards perfection. This side held a universal, timeless vision of nature, which thereby included the entire cosmos and included all of time. On the other side was an anti-perfectionist understanding of nature, by which humans are naturally imperfect and more likely to be destructive than peaceful. Yet this side also saw in nature a capacity to overcome humans' potential for self-annihilation, a capacity which delivers to only some humans the ability to study other humans so as to comprehend the seeming paradox whereby nature can be effectively overcome by nature. Comprehending this complex route to civil peace involves acknowledging that nature will allow the strongest possible ruler (because he, she, or it has the power of life and death over all those he, she, or it rules) to win control and to use this control to calm the destructive tendencies of those being ruled (because nature has instilled in those being ruled a fear of death and, hence, a fear of the individual or assembly who or which can, as unchallenged ruler, impose death on them). This side's vision of nature was, in this way, place-bound and time-bound, such that the success of the anti-perfectionist project of rule was and remains contingent on the careful reading of, and the careful management of, conditions and circumstances, so as to allow the rulers to maintain control of a territory by maintaining a monopoly of the means of large-scale violence over the internal population of that territory and over any threatening external forces.

Since the eighteenth century, these two older understandings of nature have operated more in the background than the foreground. In the late eighteenth century a new understanding emerged, as discussed above, largely through aesthetics and science. It was never implicated in civil wars in the way the earlier understandings were. Of course this fact does not mean that the battle between perfectionism and anti-perfectionism had disappeared by the late eighteenth century; it had just taken on new forms, as we will show throughout this chapter.

Before we begin to describe the new forms of this battle we think it useful to take three closely related sets of cautionary remarks we issued in the first chapter, concerned with perfectionist and anti-perfectionist thinking about the environment, and apply them to perfectionist and anti-perfectionist thinking about nature. First, as our account of perfectionist and anti-perfectionist thinking is an ideal-type account, the different understandings of nature we will describe in this chapter as more or less perfectionist or more or less anti-perfectionist are never likely to have operated, or to operate today, as pure examples of perfectionism or pure examples of anti-perfectionism. Any thinking about nature in the eighteenth, nineteenth, and twentieth centuries can be described in the same terms we used in the first chapter to describe present-day thinking about environment and climate: such thinking was and is influenced by both perfectionism and anti-perfectionism (even if the influence was or is indirect). Second, different discourses of nature – in this chapter we are mainly focusing on aesthetic and scientific discourses – can and often do combine perfectionism and anti-perfectionism in a variety of ways, as we showed in Chapter 1 when we discussed the way science is currently being used in the ongoing debate about climate change. Third, even though we can say that aesthetic thinking about nature leans towards perfectionism while scientific thinking leans towards anti-perfectionism, we must acknowledge that we are using the formulation 'leans towards' to make clear that neither perfectionist aesthetic thinking about nature nor anti-perfectionist scientific thinking about nature ever holds a monopoly within its domain.

Keith Thomas (1983: 15), focusing mostly on aesthetics, captures some important characteristics of the new eighteenth-century understanding of nature when he talks about 'a whole cluster of changes in the way in which men and women, at all social levels, perceived and classified the natural world around them. In the process some long established dogmas about man's place in nature were discarded. New sensibilities arose towards animals, plants and landscape.' Thomas makes clear that the earliest forms of the new understanding were overwhelmingly perfectionist:

> By the late eighteenth century the appreciation of nature, and particularly wild nature, had been converted into a sort of religious act. Nature was not only beautiful; it was morally healing. The value of the wilderness was not just … [to] provide a place of privacy, an opportunity for self-examination and private reverie (which was an ancient idea); it had a more positive role, exercising a beneficent spiritual power over man … The feeling of awe, terror and exultation, once reserved for God, was gradually transposed to the expanded cosmos revealed by the astronomers and to the loftiest objects discovered by explorers on earth: mountains, oceans, deserts and tropical forests.
>
> (Thomas 1983: 260)

With our eye mostly on aesthetics, we can usefully paraphrase Thomas and speak of the development of a perfectionist 'sensibility' towards nature,[1] and with it what Isabelle Lanthier and Lawrence Olivier (1999: 65) suggest is this development's 'legitimizing set of values'.

Another pointer to the early dominance of perfectionism in the new understanding of nature can be found in Carl Becker's provocatively yet accurately titled book *The Heavenly City of the Eighteenth-Century Philosophers* (1932). Becker quotes at length from a 1775 treatise by Colin Maclaurin, Professor of Mathematics at the University of Edinburgh (*An Account of Sir Isaac Newton's Philosophical Discoveries*), in showing just how much perfectionist moral and

---

1 Thomas is of course aware of the importance of Rousseau and Humboldt in this development (Thomas 1983: 260).

religious baggage was bequeathed in the eighteenth century to those endeavours that would later become the mainstay of the scientific study of the environment:

> 'To describe the *phenomena* of nature, to explain their causes ... and to inquire into the whole constitution of the universe, is the business of natural philosophy ... the inexhausted beauty and variety of things makes it ever agreeable, new and surprising ... But natural philosophy is subservient to purposes of a higher kind, and is chiefly to be valued as it lays a sure foundation for natural religion and moral philosophy; by leading us in a satisfactory manner, to the knowledge of the Author and Governor of the universe ... We are from his works, to seek to know God, and not to pretend to mark out the scheme of his conduct, in nature, from the very deficient ideas we are able to form of the great mysterious Being ... [T]hat mighty power which prevails throughout ... and that wisdom which we see equally displayed in the exquisite structure and just motions of the greatest and subtlest parts. These, with perfect goodness, by which they are evidently directed, constitute the supreme object of the speculations of the philosopher; who ... cannot but be himself *excited and animated to correspond with the general harmony of nature.*'
>
> (Becker 1932: 62–63, quoting Maclaurin, emphases in original)

For Becker (1932: 63), 'the closing words of this passage may well be taken as a just expression of the prevailing state of mind about the middle of the eighteenth century. Obviously the disciples of Newtonian philosophy had not ceased to worship ... having denatured God, they deified nature'.

The role of aesthetics in the new understanding of nature which emerged in the late eighteenth and nineteenth century

The aesthetic appreciation of nature that developed across the late eighteenth and especially the nineteenth century drew on a number of overlapping traditions, especially the tradition of the picturesque, the tradition of the sublime, the Romantic tradition, the informal landscape or gardens tradition, and the utilitarian tradition.

The picturesque tradition

The picturesque tradition valued 'natural rather than artificial or improved objects ... The new taste for the natural world included an appreciation and even a love of wild things' (Hargrove 1989: 87). But not all nature was attributed picturesque qualities. To be picturesque, a scene had to differ significantly from the ordinary familiar surroundings of the everyday. A truly picturesque scene was framed according to the conventions of landscape painting and necessarily remained at a distance, appreciated for 'its artistic and scenic qualities of line, colour and design' (Carlson 1998: 125). William Gilpin, an English aesthete who claimed special expertise in the picturesque, observed that England would '"be more beautiful in a state of nature than in a state of cultivation"', for '"[w]herever man appears with his tools, deformity follows his steps. His spade and his plough, his hedge and his furrow, make shocking encroachments on the simplicity and elegance of landscape"' (Thomas 1983: 285, quoting Gilpin). It is not difficult to spot the traces of perfectionist morality in the thinking on display here – nature is superior to anything artificial, ordinary, or cultivated – that is, anything human-made – because it is (by nature) perfect.

Despite this, for those who adopted the picturesque tradition it was acceptable to artfully rearrange nature, or even to sacrifice it to meet the needs of art. It was not unusual for a landscape photographer in the nineteenth century, for example, seeking the picturesque, to carry an axe to open up a prospect or disguise a lack in the foreground of the scene. Such photographers did not think of themselves as vandals. In fact, many of them advocated for the preservation of wild nature and the creation of national parks. The early photographer of the Tasmanian wilderness (then, as now, regarded as one of the world's most important wilderness areas) J.W. Beattie, for example, employed a clear distinction between the tools he used to 'remedy faulty composition' (Bonyhady 2000: 210), and those which he feared would come with the 'tide of utilitarianism' to sweep away nature's 'glories' (Hutton and Connors 1999: 76).

By the latter part of the nineteenth century, the picturesque ideal of natural beauty was well established in other parts of Australia besides Tasmania. James Smith, a leader writer for Melbourne's *Argus* newspaper, as well as being its art and theatre critic, built a strong case for the preservation and extension of national parks, insisting on the picturesque aesthetic of places such as Fern Tree Gully in the Dandenong Ranges in Victoria. To Smith's mind, conservation of places of natural beauty was vital to the maintenance of a population of 'civilised people':

> 'A certain portion of woodland is … essential … to the beauty of the landscape. It is the main and, indeed, in most of our districts, the sole ingredient of the picturesque, and we should be deficient in the ordinary good taste supposed to be characteristic of a civilised people, if we were insensible to the necessity of improved care and judgement in protecting and enlarging our reserves of the kind.'
>
> (Bonyhady 2000: 109, quoting Smith)

Eccelston Du Faur, another Australian colonist who sought to combine an artistic sensibility with a commitment to wilderness preservation, was active in the 1870s. He campaigned for the establishment of the Ku-ring-gai Chase National Park in New South Wales, not far from central Sydney, and later became its first Managing Trustee (Bonyhady 2000: 194). He also wanted to popularise the natural beauty of the Grose Valley in the Blue Mountains to the west of Sydney. To achieve this objective, in 1875 he commissioned the photographer Sydney Joseph Bischoff to capture the valley and its cliffs after the fashion of Carleton Watkins's internationally acclaimed images of the Yosemite Valley in the USA. To assist Bischoff, Du Faur, adopting a picturesque spirit very similar to that of Beattie in Tasmania, hired axemen and led a party of private school boys to chop away trees and bush, so that '"the abruptness of the cliffs" could "fairly be appreciated"' (Bonyhady 2000: 196, quoting Bischoff). He lamented that with more time he might have '"opened up the sky ... all around the compass, forming a view of cliffs of 2000 feet high all around, not easily to be surpassed"' (Bonyhady 2000: 198, quoting Bischoff).

When regarding the Australian landscape, some of those dedicated to the picturesque could not fully appreciate the new-to-European-eyes sights before them. They found the landscape monotonous. These doubters could not comprehend the indigenous trees, as is evidenced by this contemporary account of the Australian eucalypt in the English newspaper *The Times*:

> 'Neither so delicate nor so umbrageous as the trees of Europe: they are not so well adapted to the beauties of landscape as the oak, the elm, the beech, and the poplar of this country; there is moreover, a sameness of appearance about them, which deprives the representation of interest.'
> (Bonyhady 2000: 92, quoting *The Times*, 29 June 1835)

This was not the case, however, with John Glover, who went on to become famous for his capacity to paint the supposedly alien landscape in a way that was pleasing to most within the picturesque camp. He was full of 'the expectation of finding a beautiful new world – new landscapes, new trees, new flowers, new animals, new birds' (Bonyhady 2000: 92). He managed to make the colonial landscape pleasing to all eyes by having it conform to established (English) habits of taste. Bonyhady observes that Glover's painting 'A View in Mills Plains' combines 'everything he admired in his new surroundings, celebrating a particularly majestic tree, its park-like surrounds and its immediate suitability for grazing' (Bonyhady 2000: 91). Glover, along with many of his fellow colonists, had a preference for the more open and potentially productive landscape, which was found west of Sydney and in parts of Tasmania (then known as Van Deimen's Land). This, to him, was country akin to the parkland estates of the English gentry and was more aesthetically pleasing than the dense bush around Port Jackson in Sydney (Bonyhady 2000: 71–80).

The sublime tradition

Nature as sublime was not necessarily beautiful; indeed, it could 'be harmful as well as pleasing'. The sublime, this is to say, was not at all the same as the picturesque. The sublime was more challenging, it did not share the picturesque's sense of completeness, and it required 'a reappraisal of man's importance and position in the world' (Hargrove 1989: 87).

An excellent example of this style of the aesthetic appreciation of nature can be found in the attitudes and practices of the American John Muir, preservationist, founding president of the Sierra Club, and activist for the establishment of the Yosemite National Park. Muir embodied the tension between the picturesque and the sublime. In an account of one of his expeditions in the Sierras in the 1870s, Muir describes guiding two artists through the mountains to find a 'landscape suitable' for the paintings they wished to attempt. Muir knew just the picturesque qualities the painters were looking for. As he noted in his journal, it had to be a scene which could be separated into 'artistic bits capable of being made into warm, sympathetic lovable pictures with appreciable humanity in them' (Muir 1998: 109). It took two days of trekking for Muir to bring the artists to a satisfactory location, in this case a 'typical alpine landscape' (Muir 1998: 110).

But the two artists' response to the Sierra, his journal makes clear, was nothing like his own. For Muir, the Sierras are sublime, not picturesque:

> Standing here in the deep, brooding silence all the wilderness seems motionless, as if the work of creation were done. But in the midst of this outer steadfastness we know there is an incessant motion and change ... Here are the roots of all life of the valleys, and here more simply than elsewhere is the eternal flux of nature manifested ... [W]e also learn that as these [mountains] we now behold have succeeded those of the preglacial age, so they in turn are withering and vanishing to be succeeded by others yet born.
>
> (Muir 1998: 113)

Muir thought the sublime quality of this landscape could provide individuals with a clearer sense of the relationship between humans and nature. Nature's perfection can, if properly understood, help individual humans to improve themselves, pushing them towards perfection, albeit a perfection they can never hope (and should never hope) to emulate:

> I pushed on southward toward a group of savage peaks that stand guard about Ritter on the north and west, groping my way; and dealing instinctively with every obstacle as it presented itself ... In so wild and so beautiful a region was spent my first day, every sight and sound and inspiration, leading one far outside of himself, yet feeding and building up his individuality.
>
> (Muir 1998: 111)

The Romantic tradition

While the Romantic movement of the nineteenth century touched art and music, the romance of nature itself was perhaps more clearly expressed in literature. This was in part because Romantic literature cast nature anthropomorphically – nature was primarily understood through its capacity to 'validate' a sense of being civilized. In this sense, it followed the thinking of Rousseau, who 'recognized that we need the conception of the state of nature in order to have a critical understanding of the nature of civilization' (Bate 1991: 56–57). In line with this, Romantic literature became a means by which the 'complexity and wholeness of the environment' was represented in human terms (Taylor 2004: 134).

The Romantic literature to which we refer was pitched at a large audience, whether it took a demotic or high-brow form. At the demotic end of the scale adventure novels and travelogues re-created the bountiful qualities of nature through an imaginative exploration of wilderness and other exotic landscapes, which had been detailed and celebrated in the reports of scientific expeditions by botanists and geographers since the voyages of James Cook and Joseph Banks (Taylor 2004: 134). And after 1859 it was not unusual for a fictional narrative, such as Edith Nesbit's *Island of the Nine Whirlpools* (1899), to replicate the 'detail and the close description of island forms and environs' found in Charles Darwin's account of the *Beagle* voyages, undertaken in the period 1832 to 1836 (Taylor 2004: 129).

At the high-brow end were figures like Henry David Thoreau in America and William Hazlitt and William Wordsworth in England. All three explored and articulated the 'complexity and wholeness of the environment' by describing their own mundane experience of nature as it surrounded them daily (Bate 1991: 90). As Jonathan Bate observes, the attitude to nature typified by these writers values the 'economy of nature' for its place in the human 'social and psychological economy' (Bate 1991: 7).

Thoreau's intimate account of his wanderings and life within the very specific locale of Walden Pond in Massachusetts, first published in 1854, was very popular. In this book Thoreau is animated more by the bog than any sublime landscape. His wanderings and writings about his local area continued after his time at Walden Pond, mostly in journal form. The journals contain both his observations and his account of the relationship and responsibilities of humans to 'the actual condition of the place where we dwell' (Worster 1994: 66; Thoreau 1998).

In a similar vein, Hazlitt argued in his essay 'On the Love of Country' that nature is our 'universal home':

> 'It is because natural objects have been associated with the sports of our childhood, with air and exercise, with our feelings in solitude, when the mind takes the strongest hold of things, and clings with the fondest interest to whatever strikes its attention; with change of place, the pursuit of new scenes, and thoughts of distant friends: it is because they have surrounded us in almost all situations ... because they have been the chief source of nourishment of our feelings, and a part of our being, that we love them as we do ourselves.'
>
> (Bate 1991: 52–53, quoting Hazlitt)

In 1810 Wordsworth wrote an introduction for Joseph Wilkinson's *Select Views in Cumberland, Westmoreland, and Lancashire*. Wordsworth's main focus was on the Lakes of the area. In 1820 he expanded this piece into an essay, 'A Topographical Description of the Country of the Lakes, in the North of England', to accompany a volume of his sonnets and poems. Both versions were well received. So much so that in 1822 Wordsworth expanded them into a book, *A Description of the Scenery of the Lakes in the North of England Guide*. This was an outstanding success, particularly because of his fame as a poet. In the third edition, which appeared in 1842, the publisher went so far as to include Wordsworth's name in the title: *A Complete Guide to the Lakes, Comprising Minute Directions for the Tourist, with Mr Wordsworth's description of the Scenery of the Country, etc. And Three Letters of the Geology of the Lake District, by the Rev. Professor Sedgwick, Edited by the Publisher*. This version, often known simply as *The Guide*, was then treated as a separate book, which had five editions of its own over the following seventeen years (Bate 1991: 43). Each new edition included extra material from Wordsworth and his sister Dorothy, as well as from Sedgwick. It was only nominally a guide for tourists, with '"the humble and tedious Task of supplying the Tourist with directions"' relegated to a preface, which the publisher had to collate himself (Bate 1991: 43, quoting from *The Guide*). Its main purpose was more complex.

While serving as a practically focused, readily accessible account of the Lakes, *The Guide* also sought to demonstrate the harmony of human society, human work, and nature. Bate observes that, unlike most travellers' guides of the period, which were written 'exclusively for *visitors* to the Lakes, Wordsworth aimed to show what it meant to *dwell* there' (Bate 1991: 45, emphases in original). Indeed, Wordsworth was opposed to the Kendal–Windermere railway because, he argued, the mass of day tourists could not, without an education, appreciate the true natural and unimproved beauty of the Lakes in all their wildness (Thomas 1983: 267). Wordsworth intended *The Guide* to form a major part of this 'education'. In it he not only described the wildness and natural beauty of the district, he venerated the agricultural traditions and the historic legacy of the built environment. He observed that each of these elements '"in their very form call to mind the processes of Nature"' (Bate 1991: 47, quoting Wordsworth).

After the 1842 edition, which was dominated by Wordsworth's Romantic sensibility, *The Guide* found more room for science (Bate 1991: 42–46). As well as the regular contributions by the noted geologist Adam Sedgwick, later editions included a botanical table compiled by Thomas Gough, which listed 250 species of plants and which identified for the reader the wood or fell where they might be located. Bate argues that Sedgwick's account of the geological formation of Cumberland and its fossil record served to underwrite Wordsworth's own account of the Lakes. Sedgwick reflected, for instance, upon the manner in which the sustainability of the vales and lakes was dependent upon the presence and structure of the surrounding mountains and tarns (Bate 1991: 46).

But while it allowed more room for science, *The Guide*, as Bate notes, continued to serve primarily as an aesthetic polemic, directed to '"the Minds of Persons of taste, and feeling for Landscape"' (Bate 1991: 43, quoting *The Guide*). Wordsworth wanted his readers to use their enjoyment of the Lakes as a platform from which to raise their own consciousness, particularly on issues like the introduction of larch plantations and similar developments in the Lakes (Bate 1991: 45–46). In line with this point, Bate argues that even though it may have been forced to allow science onto the stage, in this period aesthetics still dominated the new understanding of nature, as a constant witness to the very essentials of being human and alive in the world, inspiring the system of National Parks in America and the preservation of parts of rural England so that they might be enjoyed and experienced by those who must live in 'grim, "grey homes"' (Bate 1991: 53, quoting Wordsworth).

The significance of nature, this line of thinking proposes, extends beyond the natural world. Wordsworth in particular, as we hinted earlier, wanted to preserve the integrity of the simple built environment of the Lake District and the organic qualities of its cottages, which he said '"appear to be received into the bosom of the living principle of things"' (Bate 1991: 47, quoting Wordsworth). Similarly, William Morris and John Ruskin envisioned an approach to town planning and architecture which would allow the built environment to mirror the beauty and form of nature as a whole (Bate 1991: 55). The carefully crafted domestic interiors created in Morris's workshop held the same aesthetic aspiration, as did Ruskin's and Morris's championing of the necessary dignity of work and the inherent nobleness of the artisan's craft.

For these reformers, the destruction of the natural environment reflected the destruction of humanity embodied in the crushed vitality of the children who worked the mills of industrialized northern England, a theme so well captured by Wordsworth's *The Excursion*.[2] Here, the child at work in the cotton mill

> Performs its functions; rarely competent
> To impress a vivid feeling on the mind
> Of what there is delightful in the breeze,
> The gentle visitations of the sun,
> Or lapse of liquid element – by hand,
> Or foot, or lip, in summer's warmth – perceived

(Wordsworth 1888)

In all this, we suggest, the Romantic aesthetic was a nineteenth-century vessel carrying the Kantian version of perfectionism forward.

---

2 Bate notes that 'by the mid-1830s *The Excursion* had been printed four times and sold more copies than any other volume of Wordsworth's poetry' (1991: 41).

The informal landscape or gardens tradition

If the picturesque, sublime, and Romantic traditions were overwhelmingly perfectionist in their appreciation of nature, the informal landscape tradition was one in which perfectionism was forced to contend with levels of human imperfection which the other traditions were mostly able to ignore.

Keith Thomas suggests (1983: 261–62) that the fashion for informal landscaping, which took particular hold in England in the nineteenth century, was a 'self-conscious reaction' against the enclosed and regular fields of that country's agriculture. It was inspired by a passion for unspoilt mountains and wild places (Thomas 1983: 264). In this sense, this tradition was perfectionist in something like the way in which the sublime and Romantic traditions were perfectionist. But inasmuch as these informal landscapes had to be physically constructed as gardens (usually by workers hired by those wealthy enough to afford them), they were bound to be imperfect human re-creations of perfect nature.

The participants in this garden movement did their best to keep the imperfect elements to a minimum: 'Since each individual plant was an emissary, so to speak, from some mysterious and bizarre corner of the Earth, the thoughts of the garden enthusiast inevitably turned to the contemplation of the natural and alien environment' (Hargrove 1989: 83). They very much wanted to believe that their gardens promoted the 'natural qualities and characteristics of the plants' and that each plant was something more than an object to be artfully positioned or skilfully shaped to enhance the overall design of a formal garden. This is to say that the participants promoted their plants as 'self-contained and self-organizing entities worthy of admiration and study for their own sake' (Hargrove 1989: 83–84), even if, as was often the case, they had actually taken specimens from their natural habitat (Bonyhady 2000: 114).

The utilitarian tradition

The utilitarian tradition was not concerned with attempting to perfect humans, nor was it especially interested in whether nature is perfect. As its name suggests, this late nineteenth/early twentieth-century movement had only practical goals, goals to do with finding a way for nature to directly benefit humans. As such, one can see in this tradition a partial forerunner of what would become the imperative of the legal-political government of the environment and climate change in the twentieth and twenty-first centuries: to control, as much as possible, the ever shifting balance between the benefit of humans (towards as wide an appreciable spread of civil peace as possible) and the maintenance of nature.

For the main protagonists of the utilitarian tradition, the type of value that attaches to nature is not constant or isolated from wider debates about the value of property and its appropriate use. They were not opposed to the preservation of wilderness, but they thought that preserving it for the 'aesthetic delight triggered in tourists by natural objects' (Hargrove 1989: 80) was as good a justification for preservation as any of those generated by people with a supposedly 'higher' aesthetic sensibility. And, it has to be said, if simply encouraging an interest in nature is a criterion of success, the numbers were on their side. For example, 'By the 1870s there was a torrent of published ... guides to the beauties of England, embellished by aquatints of picturesque views from 1775 and steel engravings from 1810' (Thomas 1983: 266).

The situation was no different in Australia. While aesthetics had inspired bushwalkers to successfully lobby for the creation of Australian national parks – as sanctuaries not only for native flora and fauna but also as places 'where people might find their true selves' (Hutton and Connors 1999: 61) – between the 1880s and 1920s each of the Australian states also developed rail-lines from their main cities to natural beauty spots in previously difficult-to-access places, such as the Blue Mountains and Fern Tree Gully. Cheap day-return tickets made visits to these places possible for working people, where they had previously been the sole preserve of the middle class (Hutton and Connors 1999: 64). This set up an occasional double standard regarding what was a 'proper' appreciation of nature. The Melbourne press, for example, excoriated those working-class day-excursionists who demonstrated their appreciation by collecting and taking home great bundles of wild flowers and ferns. Yet the same press turned a blind eye when the middle-class ladies of Melbourne did the same, excusing them because their 'simple trusting minds' believed the ferns to have 'been produced for the ornamentation of suburban drawing rooms' (Bonyhady 2000: 119).

The utilitarian arguments of the period extended beyond such themes as the temporary provision of a healthy and stimulating natural sanctuary from the corruption of the city (Hargrove 1989: 80; Hutton and Connors 1999: 76–77; Bonyhady 2000: 115–19). The utilitarian appreciation also began to explore, albeit in a limited way, the economic value of natural resources. For example, political pressure to relieve the problem of unemployment in Melbourne during the depression of the 1890s led authorities to open up tracts of the pristine Dandenong Ranges, dividing these tracts into five- and ten-acre blocks so that working men with little capital might establish market gardens and orchards (Bonyhady 2000: 119–23). This type of late nineteenth-century utilitarian thinking about nature established a division that would become a mainstay of late twentieth/early twenty-first-century debate about the legal-political government of the environment and climate change, a division between, on the one hand, those who hold that nature's perfection must be preserved without consideration of any value which might attach to it as an economic resource and, on the other, those who hold that to preserve nature as an economic resource is still preserving it, albeit with its perfection now to be understood in terms of its contribution to the welfare of humans.

The overlap between aesthetics and science in the nineteenth century

In the nineteenth century most scientists conducted the great bulk of their studies of nature in a manner very similar to that employed by those whose goal was pure aesthetic appreciation. Both groups relied heavily on careful observation. In this way, geologists, botanists, and phytogeographers (those who study the geographical distribution of plants) ventured into the field in much the same way as did painters, poets, travel writers, photographers, collectors, horticulturalists, and nature tourists. During the period in question the two groups comprised mostly amateurs, pursuing a respectable pastime for the love of it (Hutton and Connors 1999: 28). In this way, appreciation of nature – whether scientific or aesthetic or, as was often the case, both – became a hobby and a passion of many members of the middle class. As such, it is hardly surprising that both groups tended to adopt the same attitude towards nature's strengths and frailties. As Eugene Hargrove suggests, 'western aesthetic and scientific attitudes' were inexorably intertwined as they marched together towards an 'environmental perspective' (Hargrove 1989: ix).

This was true even when colonial expansion took the naturalists' quest into exotic territory, beyond Europe and North America. Early horticultural societies, for instance, took up the opportunity to work alongside the natural scientists who were collecting seeds and specimens from these locations. The specimens were then showcased in the landscaped public and private gardens of Europe, America, and even Australia,[3] where they were readily available for further study and appreciation (Hargrove 1989: 82–83).

A useful way of illustrating the overlap between aesthetics and science is to consider the manner in which the two endeavours sometimes ended up being practised by the one person. Thoreau, for instance, was a 'self educated naturalist, a competent field ecologist' (Worster 1994: 60).[4] He wrote papers and gave lectures on seeds and on 'The Succession of Forest Trees' (Worster 1994: 69–71). Thoreau had read Linnaeus closely, but he did more than learn the species of the plants and animals he studied. Inspired by the writings of Lyell, Darwin, and Humboldt, he tried to develop a more holistic understanding of the interrelation between the various elements of the animate and inanimate environment in which he lived (Worster 1994: 65).

---

3 The gardens of John and Elizabeth Macarthur's 'Camden Park', near Sydney, were designed and planted by their son William in the 1820s. A passion for botany had led him to become part of the world-wide exchange of exotic botanical specimens. He both collected specimens himself and sponsored others to do so, such as the explorer and naturalist Ludwig Leichardt (Kerr-Forsyth 2005: 64).

4 Obviously, given our earlier discussion of the *OED*'s definitions of the term 'ecology', we do not think Thoreau can sensibly be called an 'ecologist', though we acknowledge that there are grounds for Worster to think so, inasmuch as the *OED* offers a note to the effect that 'A supposed use of ecology in a letter by Thoreau dated 1858 represents a misreading of geology' (*OED* online).

The overlap between the two groups also had a more pragmatic side, inasmuch as scientists had come to rely on artists and artists had come to rely on scientists. Before the development of reliable photography, scientists needed artists with them in the field to record the natural phenomena they were studying. As Hutton and Connors argue (1999: 28), the collecting and studying of the scientist and the illustrating of the artist were remarkably similar operations, so it is not surprising that 'the attention to detail required by both was often accompanied by [a shared] sensitive appreciation of the natural landscape'. Hargrove notes (1989: 85) that the seminal scientist Alexander von Humboldt included a chapter in his *Cosmos: A Sketch of a Physical Description of the Universe* (published in 1852) on the vital role of 'landscape painting in the study of natural history'. On the other side of the coin, by the middle of the nineteenth century artists, partly through the influence of critics like Ruskin, were seeking to paint landscapes with a degree of scientific accuracy (Hargrove 1989: 85). Ruskin's *Modern Painters* stressed the need for a connection between artistic techniques and a scientific frame of mind. Ruskin's work impressed the American landscape painter Frederick Edwin Church, who in the late 1850s, tracing one of Humboldt's expeditions, set out to 'paint the natural history of South America'. While he was in Ecuador, Church painted *The Heart of the Andes* (1859), which, with its close attention to the actual forms of nature, was to establish in mainstream American landscape painting a convention of scientific verisimilitude (Hargrove 1989: 85).[5]

Before we head into the next section, it is worth stressing that even within the terms of the overlap between the aesthetic and the scientific, there were significant differences. Bate argues that when important contributors to the aesthetic appreciation of nature, such as Wordsworth and Thoreau, were borrowing from science, they were still a long way from the thinking of scientists like Darwin and Haeckel: 'Scientists made it their business to describe the intricate economy of nature; Romantics made it theirs to teach human beings to how to live as part of it' (Bate 1991: 40).

The role of science in the new understanding of nature which emerged in the late eighteenth and nineteenth century

Methods, tools, background conditions, the work of particular individuals, the work of particular disciplines, and, perversely perhaps, the continuing preoccupation with morality were all important factors in science playing the part it did in creating and promoting the new understanding of nature.

5   In a sort of feedback loop, scientists started bringing questions of artistic taste to their work. As Hargrove notes (1989: 84), when in the field natural scientists often 'matter-of-factly jotted down their aesthetic judgments alongside their factual descriptions'.

The importance of methods, tools, and the right background conditions

From the seventeenth through to the nineteenth century, under the cover-all term 'natural history', the enthusiasm for scientific inquiry 'provided so to speak, a universal language for ordering form; taxonomical methods occupied minds on both sides of the English Channel' (Taylor 2004: 24). The 'universal language' was largely concerned with observation and experimentation. As Laudan (2003: 593) puts it, the 'new science' that emerged in the second half of the seventeenth century was built around work in the laboratory, work which aimed to test or demonstrate scientific theories through careful observation and experimentation.

The idea that *scientific* observation and experimentation was a special, higher form of observation and experiment was central to the establishment of learned scientific societies, like the Royal Society of London, which, as we mentioned earlier, was established in 1660. Like other learned societies, the Royal Society served as a gate-keeper, a means of deciding what was 'proper' science and who were 'proper' scientists. It also served to promote active engagement with science among the wider population, as a practical art and technology (Pyenson and Sheets-Pyenson 1999: 75, 80). In this way, the culture of natural-scientific inquiry 'ceased to be solitary and introspective … it became shared and communal'; each scientific society became a 'vital instrument for formulating and transmitting scientific norms and values' (Pyenson and Sheets-Pyenson 1999: 74–75).

Across the course of the eighteenth century the techniques of observation and experimentation became increasingly sophisticated, a process which both led to and was led by the invention of new tools, like the microscope, telescope, composite lenses, pendulum, and air pump (Laudan 2003: 593). More and more of the natural world fell under scientific scrutiny, such that nature soon came to be understood 'as a machine-like entity subject to disaggregation' (McMichael 2001: 160). The progression of the 'new science' was also aided by mercantile and colonial expansion, which allowed the ready transmission of correspondence, documents, details, observations, and reports between individuals and societies of similar scientific mind but located in different countries (Pyenson and Sheets-Pyenson 1999: 88–90).

The importance of particular individuals

In 1627 Francis Bacon, one of the earliest polemicists for the new scientific project, argued for the formation of more and more scientific societies, saying they were vital to the task of delineating '"the knowledge of causes, and the secret motions of things; and the enlarging of the bounds of Human Empire, to the effecting of all things possible"' (Pyenson and Sheets-Pyenson 1999: 77, quoting Bacon).

Humboldt did something similar in the late eighteenth and early nineteenth century. Following the successes of Cook and Banks's voyage to the

South Pacific, which served to stoke the 'mania for scientific exploring' (Pyenson and Sheets-Pyenson 1999: 255), Humboldt became the model scientific explorer. His expeditions were heroic in scale, exotic in their destination, and employed the observational techniques underpinning experimental work in the laboratory. His aim was to measure and chart the juxtaposition of the physical forces of the natural world, what he called 'terrestrial physics' (Dettelbach 1996: 289–300). He collected at least 6300 new plant specimens during one long tour of the Americas between 1798 and 1804 (Pyenson and Sheets-Pyenson 1999: 258). More than this, he categorized every specimen according to the type of environment in which it was found, and on the basis of this work he 'divided the globe into discrete vegetation zones', which allowed him to describe the similarities in plant species found in equivalent environmental or vegetation zones (Cittadino 2003: 230). Through his calculations, cataloguing, and collections (and those of others like him), discrete natural habitats were rendered visible for the first time.[6]

The role of phytogeography and biology

This remarkable explosion of interest in the natural world took many forms; in particular nature was explored and charted by phytogeographers (and, as we saw in the previous section, it became a focus for horticultural enthusiasm, for adventure, for contemplation, for spiritual renewal, and for travelogues, novels, and poetry). By the beginning of the Victorian period in England 'natural history and field botany' had become a 'characteristic … feature of provincial life' (Thomas 1983: 283).[7] By the close of the nineteenth century many of the components of what would later become the study of ecology/environment were in place. There was 'a rich collection of observations and descriptions', as well as 'a growing vocabulary of phytogeographical terms' and a growing contextual catalogue of exotic organisms (Cittadino 2003: 230). In this way, the nineteenth-century phytogeographers provided the inspiration for the emergence of modern ecological science (Boughey 1971: 5).

Of course other disciplines besides phytogeography made contributions at least as important, if not more so. Probably the most important of all was biology, which emerged as a distinct discipline when, in the nineteenth century, the life sciences developed and expanded to the point that practitioners began forming professional disciplinary associations of their own and working in specialist research laboratories (Pyenson and Sheets-Pyenson 1999: 97).[8]

---

6 There was a negative side to all this: the prodigious collecting and cataloguing of botanical specimens was sometimes to the detriment of favoured species and sometimes entailed the degradation of places of outstanding natural beauty (Bonyhady 2000: 109–13).

7 The situation in Australia was similar. By the late nineteenth century field naturalists had established societies in many provincial towns as well as all the major cities (Hutton and Connors 1999: 31).

8 As Laudan notes (2003: 593), the emergence of biology was dependent upon the 'progressive extension of experimentation', which facilitated 'the manipulation of nature in controlled settings'.

The rise of biology owed a great deal to advances in bacteriology. 'From its earliest dramatic achievements, bacteriology was an icon of the triumph of technical skill and experimental discipline over speculation and superstition' (Mendelsohn 2003: 77). In trumping miasmic theory, bacteriology made individuals the site of identifiable toxic exposure. Consequently, particular microbiological agents were linked to the onset of various diseases, rather than to the vaguer pathogenic environment. 'It was now the details of the foreground rather than the larger constructs in the background that commanded attention' (McMichael 2001: 163). This is to say that, 'Bacteriology gave medicine and hygiene powerful reasons for focusing on the identification and control of necessary, specific causes' (Mendelsohn 2003: 77).

Microbiologists like Koch and Pasteur effectively entrenched the reductionist focus and method of scientific positivism, their significant experimental breakthroughs having expanded the breadth and depth of biological understanding, which furthered scientific specialization (McMichael 2001: 163). Biological science, this is to say, did not follow a simple unified pattern of development, rather, it had antecedents in a wide array of 'research traditions and practices'. These 'traditions and practices' included midwifery, medicine, forestry, agriculture, biotechnology, geography, breeding, collecting, experiment, field work, voyages of exploration, dissection, and taxonomy. Unlike the procedure that dominated the physical sciences, by which new discoveries usually meant abandoning existing ways of doing things, in biology old research practices more readily continued alongside new approaches (Pancaldi 2003: 92).

In line with this, it is fair to say that as a science, modern biology, largely by the breadth and depth of its focus and its methods, eventually rendered the 'observational and anecdotal study of natural history' unscientific (Boughey 1971: 3). The development of cellular biology confirmed and hurried this process. Schleiden's work on plant cells and Schwann's work on animal cells led to 'the cell' being classified as the fundamental unit of life. Innovations in techniques and instruments, such as the compound microscope, allowed the dissection of cells and the cultivation, manipulation, and representation of micro-organisms (Pancaldi 2003: 93). From the laboratories, especially those of Koch, Pasteur, and Lister, came a new understanding of the involvement of microscopic forms of life 'in the processes of putrefaction and disease' (Mendelsohn 2003: 76).

The moral dimension

As we stressed in the introduction to this chapter and in our discussion of the role of aesthetics, morality was rarely far from the concerns of the institutions, individuals, and practices which contributed to the important role played by science in fostering the new understanding of nature. For example, in 1854 John Hutton Balfour, a Professor of Botany, claimed that the '"examination of plants in their living state, and in their native localities" would combine "the pursuit of scientific knowledge" with the "healthful and spirit-stirring reaction which tends materially to aid mental efforts"' (Taylor 2004: 126, quoting Balfour).

In 1860 Alfred Wallace discoursed at length on birds of paradise and in doing so ended up expressing a sentiment that would later become a mainstay of environmental politics – the near-impossibility of reconciling humans' quest for 'civilization' and their moral duty to refrain from disturbing the natural world, in which they understand themselves to be intruders:

> 'It seems sad that on the one hand such exquisite creatures should live out their lives and exhibit their charms only in these wild inhospitable regions ... while on the other hand, should civilized man ever reach these distant lands, and bring moral, intellectual and physical light in to the recesses of these virgin forests, we may be sure we will disturb the nicely-balanced relations of organic and inorganic nature as to cause the disappearance, and finally the extinction, of these very beings whose wonderful structure and beauty he alone is fitted to appreciate and enjoy. This consideration must surely tell us that all living things were not made for man. Many of them have no relation to him. The cycle of their existence has gone on independently of his, and is disturbed or broken by every advance in man's intellectual development.'
>
> (McCalman 2009: 279, quoting Wallace)

As Thomas puts it, the 'intense interest in the natural world' of the eighteenth and nineteenth centuries developed concurrently with 'doubts and anxieties about man's relationship to it which we have inherited in magnified form' (Thomas 1983: 15). The nineteenth-century debate about the importance of geological time, which became part of the battle over evolution, provides a fascinating example of Thomas's point. While allowing that divine creation had set the machinery of geological change in motion, the famous geologist Charles Lyell expounded a set of 'professional scientific principles' to explain how the 'slow everyday forces of wind, rain, ice and sun' had transformed and shaped the earth's geological form over time. As a consequence (a partly unintended one), he 'freed the study [of geology and of nature] from biblical time scales' (McCalman 2009: 54). As Darwin himself summed up Lyell's achievement, '"it altered the whole tone of one's mind, and therefore ... when seeing a thing never seen by Lyell, one yet saw it partially through his eyes"' (McCalman 2009: 55, quoting Darwin).

For other nineteenth-century thinkers, such as Herbert Spencer, the debate about evolution demonstrated the need for a new moral doctrine, concerned with the 'survival of the fittest' amongst humans. By this doctrine strong humans become stronger by dominating nature, not by leaving it alone. Alfred Wallace took a similar line. In his autobiography, published in 1905, he recounted his thought process when he had been on an expedition to the Moluccas and New Guinea in 1858 and was suffering from malaria.

Stimulated by the significance of Malthus's *Principles of Population* to the study of evolutionary processes,

> 'I thought of ... "the positive checks to increase" – disease, accidents, war, and famine – which keeps down the population of savage races to so much lower an average than that of more civilized peoples. It then occurred to me that these causes or their equivalents are continually acting in the case of animals also ... [T]he destruction every year from these causes must be enormous in order to keep down the numbers of each species ... [I]t occurred to me to ask the question, why do some die and some live? And the answer was clearly that on the whole the best-fitted live ... Then it suddenly flashed upon me that this self-acting process would necessarily *improve the race*, because in every generation the inferior would inevitably be killed off and the superior would remain – that is, *the fittest would survive*.'
> (McCalman 2009: 285–86, quoting Wallace, emphases in original)

For Darwin himself the situation was more nuanced. While he had earlier gleaned from Malthus a message not dissimilar to that expressed by Wallace – especially during 'a time when the mass bread riots and hunger marches were also reminding him of the struggle for existence' (McCalman 2009: 287–300) – he later decided that the new sciences, especially biology, revealed an increasingly complex natural world to which humans were inextricably connected but over which they had little control (McMichael 2001: 20).

The battle over evolution was of course the source of many 'moral lessons', some of them extreme, some not so extreme. One of the more celebrated moments in the battle occurred at 'a meeting of the British Association for the Advancement of Science in Oxford in late June 1860', attended by up to a thousand people. At this meeting the Bishop of Oxford, Samuel Wilberforce, made a 'light and witty attack on Darwinian evolution' and then pointedly provoked one particular member of the audience, Thomas Huxley, who was an ardent supporter of Darwin's. Wilberforce asked Huxley 'whether it was on his grandmother's or grandfather's side that he was related to an ape'. Huxley replied that he would rather '"have a miserable ape for a grandfather" than a man "possessed of great means of influence & yet who employs ... that influence for the mere purpose of introducing ridicule into a grave scientific discussion"' (McCalman 2009: 344, quoting Huxley).

Joseph Hooker, another Darwin supporter, was also present. He wrote to Darwin to describe the atmosphere:

> Well Sam Oxon [Wilberforce] got up and spouted for half an hour with inimitable spirit, ugliness, emptiness and unfairness ... he ridiculed you badly and Huxley savagely. Huxley answered admirably and turned the tables, but he could not throw his voice over so large an assembly, nor command the audience; and he did not allude to Sam's weak points nor put the matter in a form that carried the audience. The battle waxed hot

... the excitement increased as others spoke; my blood boiled ... I swore to myself that I would smite the Amelekite Sam, hip and thigh if my heart jumped out of my mouth ... [T]here and then I smashed him amid rounds of applause. I hit him in the wind at the first shot in ten words taken from his ugly mouth; and then proceeded to demonstrate in as few more: (1) that he could never have read your book, and (2) that he was absolutely ignorant of the rudiments of Bot[anical] Science ... Sam was shut up and had not a word to say in reply, and the meeting was dissolved forthwith, leaving you master of the field after four hours of battle.'

(McCalman 2009: 345, quoting Hooker)

Conclusion

In the vital debate described immediately above, in which Wilberforce was defending what turned out to be the fading forces of creationist science, while Huxley and Hooker were defending Darwinism, which soon turned out to be the dominant force in biology, we see not just the 'Tory churchman versus freethinker, age versus youth, creation science versus evolution, church versus science, religion versus reason, spirit versus matter' that McCalman sees (2009: 346). We also see an odd mix of perfectionist and anti-perfectionist elements coming together in different accounts of nature, some based on the idea of perfect nature and perfectible humans (whether in the image of perfect God or not), others based on doubts about whether nature can be considered perfect or perfectible and doubts about whether humans can have any control over nature. This combination of doubts and certitude was both the moral legacy of debates from earlier centuries and a vital moral/religious ingredient in the understanding of nature that travelled with eighteenth- and nineteenth-century aesthetic and scientific perspectives into the twentieth century, right up to the moment when the environment emerged as an object of legal-political government. Of course by that time, aesthetics (like religion) was not as visible as it had been earlier, and science had been changed dramatically.

In the first half of the twentieth century science had become more anti-perfectionist than it had ever been. While anti-perfectionist elements were of course present in the embryonic sciences of the seventeenth and eighteenth centuries – partly through the influence of the early modern revival of scientific scepticism (Hankinson 1995; Hookway 1990; Popkin 1968), partly through the influence of the early modern revival of Epicurean and Stoic attitudes to science (Osler 1991), and partly through the influence of the revival of other ancient sources – these elements gained strength considerably through the late nineteenth and especially in the early twentieth century. Particularly important in this regard was a shift in the process for distinguishing genuine science from non-science. The process moved away from the determinations of particular individuals and particular scientific societies and towards a particular process, which Karl Popper later called falsifiability. As this process, and others which had the same effect, became widely established (see especially: Popper 1959), the balance between aesthetics and science shifted very much in favour of science. Legal-political government now turns more regularly to science in dealing with the environment and climate change than it does to aesthetics. Paul Rutherford (2000) goes so far as to argue that the environment became a governmental problem because of the way it was represented by science, particularly through the environmental modelling developed by scientific ecology.

Despite this science domination, it would be unwise to think that aesthetics has been chased out of the legal-political government of the environment and climate change altogether. We invite the reader to consider the examples offered in the previous chapter – offered by way of illustrating points about the operation of law, politics, sovereignty, state, economy, and interests – and to think whether aesthetics has been chased out of any of them. We doubt that it has. Whether it is water pollution, dam building, coastal erosion, the case of individuals who work for state agencies having to put the 'person' of the state ahead of their own 'personalities', or the case of a government putting economic considerations ahead of environmental considerations, aesthetics is almost certain to be present alongside science, law, politics, sovereignty, state, economy, and interests (as well as morality and religion of course). Indeed, it might be that aesthetic considerations are so important to a decision or an action about protecting water supplies, building a dam, preventing coastal erosion, disciplining a state worker, or fostering economic growth that one can sensibly talk about the addition of 'aesthetic interests' alongside the interests of science, of the law, of the state, of the economy, of religion/morality, and of particular individuals to a proper consideration of any or all of these examples.

Having made aesthetics and science into a more central part of the fabric of our answer to our key question – how do law and politics, entwined in various complex relations with sovereignty, the state, morality, religion, economy, aesthetics, and science, attempt to govern the environment and climate change? – we feel the need to reiterate that, because the environment is a chameleon, the legal-political government of it must be a complex, multifaceted endeavour, not open to easy description or explanation.

It is to the period in the twentieth century after the environment had been put in place as an object of legal-political government that we turn in our next two chapters, presenting examples of the legal-political government in which the law is the main player – in Chapter 4, private law, in Chapter 5, public law.

4 Law's role in the legal-political government of the environment

Part 1: the private common-law role

Introduction: the sphere of law's operation

In Chapter 2 we outlined the historical development of the legal-political form of government. In Chapter 3 we brought our account of its operation up to that moment in the twentieth century when it first began to attempt to govern the environment. In both places we were not only transparently explaining the historical parts played by each of law, politics, sovereignty, state, interests, economy, nature, aesthetics, science, morality, and religion, we were also obliquely introducing the idea of a distinct sphere of operation for this way of governing per se and, by extension, this way of governing the environment and climate change. In choosing to do this obliquely, we did not name this sphere, not thinking it appropriate to do so until we came, in the present chapter, to the twentieth/twenty-first-century era of environmental government, when we must talk specifically about the two distinct modes by which law shoulders much of the task of governing this newly emerged object – the private common-law mode and the public-law statutory mode.

Our oblique dealings, in the previous two chapters, with the sphere of operation of legal-political government included distinctions such as that between unregulated private matters of faith and the state regulation of churches as public voluntary associations, that between a private religious conscience and a public legal conscience, and that between private morality and the public morality of legal-political government (a morality concerned only with fostering widespread appreciable civil peace). In leaving the oblique register behind we must now stress that the *public* side of these and other such distinctions *is* the sphere of operation of anti-perfectionist legal-political government, and is, therefore, the sphere of law's operation under this form of government. To put it the other way around, the sphere of operation for the legal-political form of government is *the public sphere*.

In making our account of this sphere transparent, we know we must be careful, for we are fully aware that the term 'public sphere' has, in much legal, political, and social scientific debate, become the province of a certain Kantian style of perfectionism. This is not in any way satisfactory for what we have to say about the private common-law mode of governing the environment (to be

presented in this chapter) or for what we have to say about the public-law statutory mode (to be presented in the next chapter). We therefore need to spend
some time now, as our first main section, differentiating our anti-perfectionist
understanding of the public sphere from one particular perfectionist-leaning
account, written by Jürgen Habermas, which has been the dominant account
of the public sphere since 1962, when Habermas first published *The Structural
Transformation of the Public Sphere* (1989; for more discussion of Habermas's
contribution, see Calhoun 1992; Melton 2003).

Once we have done this differentiating work we will be in a position to
discuss in more detail the private common-law mode of governing the environment, which we will do using the example of toxic tort. This discussion
will include a lengthy consideration of the notion of risk in the common-law
context, leaving us to say more about its application to public law in the
following chapter.

Our anti-perfectionist account of the public sphere differentiated from the dominant perfectionist account

By our account, the public sphere is precisely the sphere of appreciable civil
peace that is maintained by the work of legal-political government. That
is, for us, it is the sphere created by the process we described in detail in
Chapters 1 and 2, whereby a long period of civil war was brought to an end
such that citizens in law-and-politics countries could, and still can, live their
lives without the constant threat that they might be killed simply for their
religious/moral commitments. This is the sphere in which legal-political
government now operates, the sphere in which it conducts its very 'public'
form of rule, as evidenced by the development of public policy, public financ-
ing, public health, public welfare, and, of course, public law. We have done
enough by way of describing the legal-political form of government that
produced, maintains, and operates within this type of public sphere, so we do
not feel the need to say much more by way of explaining our account of the
way the public sphere works, for it works as part and parcel of legal-political
government. Our account of the public sphere, this is to say, goes hand-in-
glove with our account of legal-political government itself: as legal-political
government developed it created and maintained the public sphere as best
it could and as the public sphere developed it helped to foster and maintain
legal-political government as best it could. This means, of course, that, by
our account, the legal-political government of the environment and climate
change is conducted entirely within this public sphere.

But one thing we have not done yet is to explain (and criticize) the rival,
perfectionist-leaning account of the public sphere which has taken such a
strong hold of debates about the public sphere over the past fifty years.
We will begin our explanation by saying that this account is the obverse
of our anti-perfectionist legal-political account: where we earlier stressed
(at different points) that legal-political government is not government by

religion-and-morality, not government by reason-and-morality, not government by private moral conscience, and not government by lofty ideals, we can now say that the perfectionist-leaning account of the public sphere insists that the public sphere is centrally about reason and morality (which has subsumed the religion–morality nexus), about the superiority of the private moral conscience, and about lofty ideals. Where our anti-perfectionist account actively refuses the idea of a quest for morally pure law and morally pure politics, sourced in reason, the perfectionist-leaning account has this quest at its core.

The influential Habermasian version of this perfectionist-leaning account contains an Aristotelian component, a Ciceronian component, and a component born of the idea of the republic of letters, but the overriding figure behind it is Kant (we base much of our discussion of the Habermasian understanding of the public sphere on Wickham 2010).

Aristotle proposed that 'man is a political animal whose "rational and sociable" nature can be completed or perfected' in a special sphere, comprising only those capable of high-level reasoning, a sphere he termed 'the *polis*'. As Hunter puts it, this special sphere is meant to inculcate in those who ascend to it 'the virtues of benevolence, charity and mercy, culminating in the summatory virtue of justice' (Hunter 2010: 477, n. 12). This is relatively straightforward. The Ciceronian component is slightly less straightforward, involving as it does the use of the Latin term *res publica*. David Burchell argues that in Cicero's time this term was not primarily concerned with the 'republic' or with what was involved in being 'republican'. Instead, he says, the 'most primal' meaning of *res publica* was 'public affairs' and 'the capacity of at least some of the citizenry to intervene in those affairs with some effect' (Burchell 2002: 93). In this way, Burchell continues, Cicero agreed with Aristotle that the public sphere must be a 'space in which political affairs can be debated', but he wanted more. In particular he wanted an emphasis on *dignitas*, which he understood to be 'the crucial attribute of that special group of citizens who aspire to high office', entailing as it did the 'almost desperate need ... to secure a kind of immortality through their exploits' (Burchell 2002: 95–96).

The republic of letters component harks back to 'the disaffected intelligentsia' of eighteenth-century Europe, that is, to those who felt slighted by the emerging rule of the state, thinking of themselves as rejected experts in public matters, pushed off the stage by a new class of state officials, 'Since all the best and brightest in the land (in their own minds, at least) were relegated to the private sphere', they tried to revive their role as 'natural leaders' by appealing directly to the 'all-encompassing "public"', especially through 'the letter, the printed tract, the literary compendium', wanting, by the use of these means, to force the state to recognize them as the source of 'a new entity which they termed "public opinion"' (Burchell 2003: 14–15; see also Koselleck 1988; Sauter 2004).

In Chapter 2 we discussed the dominant role Kant has played since the eighteenth century in the perfectionist campaign to overturn the gains of the anti-perfectionist legal-political or civil tradition of thought, that is, the campaign to have reason and morality once again make politics, law, sovereignty, and the state subservient to them, so that this collection of forces (politics, law, sovereignty, and the state) would once again serve as humble servants to perfect nature. For Kant, perfect nature promotes the 'perfect political constitution', which he thinks must be restored to its rightful place: 'The history of the human race as a whole can be regarded as the realisation of a hidden plan of nature to bring into being an internally – and for this purpose also an externally – perfect political constitution, as the only condition in which she [nature] can fully develop all of her capacities in mankind' (Kant 1970: 50).

As we have noted at a number of points in preceding chapters, many twentieth and twenty-first-century scholars seek to continue Kant's work by employing a certain style of metaphysics to minimize the relative importance of politics, law, sovereignty, and the state. In this regard, Hunter says that, 'European university metaphysics can be characterized as an academic discipline (or culture) whose thematics concern the relation between an infinite, atemporal, self-active, world-creating intellect and a finite … worldly being' (Hunter 2006: 98; see also Hunter 2009). Politics, law, sovereignty, and the state are to be kept in place by 'tethering them to philosophical reflection on … infinite being', against which they are always found to be inadequate (Hunter 2006: 98). Kant's *Religion within the Bounds of Mere Reason* was and remains a key text for this campaign:

> Here Kant reworked the terms of his own metaphysics of morals by transposing the relation between noumenal moral being and man's finite sensuous nature into a temporal register. He treated this relation as one in which history would take place as the progressive refinement of man's sensuous nature, more or less in accordance with a Protestant account of the descramentalization and dehistoricization of religion. This would eventually lead to the appearance of beings capable of governing themselves through pure reason in the temporal sphere: the kingdom of God on earth.
>
> (Hunter 2006: 98–99)

This brings us to the matter of how Habermas expresses his debt to Kant in *The Structural Transformation of the Public Sphere*. Usually he is remarkably clear about how much his account of the public sphere owes to Kant and to Kantian thinking: 'Even before "public opinion" became established as a standard phrase in the German-speaking areas, the idea of the bourgeois public sphere attained its theoretically fully developed form with Kant's elaboration of the principle of publicity in his philosophy of right and philosophy of history' (Habermas 1989: 102). Habermas praises the eighteenth-century members of the republic of letters, who 'aimed at rationalizing politics in the name of morality … the Aristotelian tradition of philosophy of politics was reduced in a telling manner to moral philosophy, whereby the "moral" (in any event thought of as one with "nature" and "reason") also encompassed the emerging sphere of the "social"'

(Habermas 1989: 102–03). He even borrows from Kant a motto for what he himself is trying to do: '"Thus, true politics can never take a step without rendering homage to morality"' (Habermas 1989: 103, quoting Kant).

Habermas, then, is no less committed than was Kant to the idea of 'a perfectly just order'. He is confident that such an order was achieved in the eighteenth century, under Kant's guidance: 'Compulsion could then no longer occur in the form of personal rule or of violent self-assertion but only in such a fashion "that reason alone has force"' (Habermas 1989: 103, quoting Kant). At this point in the eighteenth century, he thinks, a core group of enlightened intellectuals, the 'true philosophers', with Kant taking the lead, 'formed themselves into a public and ... endowed the sphere of its critical use of reason' (Habermas 1989: 104). This is the basis of his version of the public sphere that now dominates debate in the humanities and social sciences.

For Habermas, this was (and remains), the only type of public sphere 'that could guarantee the convergence of politics and morality' (Habermas 1989: 103–04). As was the case for Kant, Habermas thinks that only those able to recognize the superiority of morality and reason will participate in this public sphere: 'With regard to the individual, this denoted a subjective maxim, namely: to think for oneself' (Habermas 1989: 104). This, he is sure, will automatically include 'scholars ... concerned with the principles of pure reason – the philosophers' (Habermas 1989: 104). They are to be trusted, he thinks, 'because', in Kant's words, 'this class is by its nature incapable of forming seditious factions or clubs, it cannot be suspected of spreading propaganda' (Kant 1996: 338).[1]

There is one issue, however, on which Habermas appears, at least at first glance, to be much more democratic in outlook about membership of the public sphere than was Kant. Where Kant thought it unlikely that non-reasoners would rise to the challenge, that membership would be restricted only to those 'within the broader bourgeois strata ... the men ... of the *salons* in which "mixed companies" engaged in critical discussions', Habermas suggests that 'anyone who understood how to use his reason in public qualified ... and by no means only the philosophers' (Habermas 1989: 104–06). But this turns out to be a case of Habermas seeking to bolster Kant's argument, not to qualify it. For example, Habermas insists that 'the principle of popular sovereignty could be realised only under the precondition of a public use of reason ... Political actions ... were themselves declared to be in agreement with ... morality only as far as their maxims were capable of, or indeed in need of, publicity' (Habermas 1989: 107–08). For Habermas (as Kantian as Kant himself), if people are to gain entry to the public sphere, they have to learn how to downplay their politics in favour of their morality: 'to begin with an abstinence from the methods of precisely that political coercion from which publicity promised liberation' (Habermas 1989: 109).

1 These 'pure philosophers', as Habermas sees their role, are vital because, 'World progress ... was in need of their unhindered activity in public' (Habermas 1989: 116).

Habermas, this is to say, always supports Kant to the hilt, taking on even his most extreme points. Here is one such point:

> Only *property-owning private people* were admitted to a public engaged in critical political debate, for [they] ... were their own masters: only they should be enfranchised to vote – admitted to the public use of reason in the exemplary sense ... [T]he propertyless were excluded from the public ... without thereby violating the principle of publicity. In this sense they were not citizens at all, but persons who with talent, industry, and luck some day might be able to attain that status.
>
> (Habermas 1989: 109–11, emphasis in original)

And here is another:

> 'He can be considered happy in any condition so long as he is aware that, if he does not reach the same level as others, the fault lies either with himself (i.e., lack of ability or serious endeavour) or with circumstances for which he cannot blame others.'
>
> (Habermas 1989: 110, quoting Kant)

Habermas's thinking on the role of law and politics in the public sphere, this is to emphasize, is fully aligned with Kant's thinking. For the Habermasian understanding, we say again, the public sphere is precisely the domain where law and politics are made to kneel before morality and reason:

> In the framework of a comprehensively norm-governed state of affairs (uniting civil constitution and eternal peace to form a 'perfectly just order') domination ... was replaced by the rule of legal norms – politics could in principle be transformed into morality ... [I]t was not difficult for Kant to suppose, within the framework of his philosophy of history, that ... *the* juridical condition [the perfectly just order] would emerge out of natural necessity ... [which allowed him] to turn politics into a question of morality ... [T]he subsumption of politics under morality ... [meant that] the public sphere ... [would] keep politics in harmony with the laws of morality.
>
> (Habermas 1989: 108–12, emphasis in original)

We trust that the reader now understands why the idea of a public sphere in which a supposedly timeless and universal morality is the dominant force is anathema to our account of the way in which law and politics attempt to govern the environment and climate change. For us, law and politics attempt to govern the environment and climate change in a public sphere, that is, in a sphere in which law and politics themselves, along with the sovereign state, are the dominant forces. While we continue to acknowledge the importance of morality as a guide to living and dying, for us this is a private matter. By our account the only morality that has a place in the public sphere is the norm that drives the combination of the sovereign state, law, and politics: the widest possible appreciable spread of civil peace.

Private common-law government of the environment: the example of toxic tort[2]

For ease of presentation, we will divide this long section into five sub-sections. We will first describe the basic components of toxic tort. Second, we will introduce the central notion of risk. Third, we will describe some of the key features of the courtroom stage of a toxic tort case. The fourth and fifth sub-sections will remain in the court setting, but focus on two particular aspects of it – risk in court (fourth sub-section) and science in court (fifth sub-section).

The basic components of toxic tort

Toxic tort is a field of private law directly concerned with particular environments and the harms they might cause humans. A toxic tort is an action initiated by an individual plaintiff. A toxic tort case might be run as a 'test case', which means that if it is successful, actions on behalf of an identified 'class' of plaintiffs might follow. Suing in toxic tort (negligence), then, is one form of legal response to the unanticipated hazards of twentieth- and twenty-first century technology and consumption. If a toxic tort suit is successful it can help improve the environment in general precisely by improving the particular environment involved in the case. The improvement can be the direct result of the case, inasmuch as the losing defendant might be forced to rectify a dangerous environment, and/or it can be the indirect result of the case, inasmuch as the damages paid to the winning plaintiff might lead others who are involved in practices which may be the subject of a future suit to change these practices to avoid this outcome.

Toxic tort, then, is the type of legal action one might wish to take if one found oneself adversely affected by a toxic environment. In such a circumstance one might regard a toxic tort suit as potentially the best way to hold an errant company, corporation, or government agency accountable and legally liable. In this way, toxic tort is a likely vehicle for a very particular sort of legal government of the environment, one made even more likely by the fact that it has been popularized by Hollywood and made practically possible by innovative law firms such as the Australian plaintiffs' firm Slater and Gordon, which has developed a powerful reputation for itself through a number of high-profile, successful toxic tort cases (Cannon 1998). What makes toxic tort even more attractive, at least superficially, is the supersized damages awarded in mass toxic tort claims in the United States (where the cases are usually 'class-actions' or 'mass torts'). As well, as Peter Huber argues in his *Galileo's*

2 Our argument in this section owes a considerable debt to Goodie (2008).

Revenge: Junk Science in the Courtroom (1991), toxic tort doctrine has allowed individuals to rationalize their misfortune through tenuous connections to improbable causes (Huber 1991). Sheila Jasanoff (1995: 12) puts the same point more sympathetically when she says that 'courts are often the first social institutions to give public voice and meaning to inaudible struggles between human communities and their technological creations'.

But the reader should not be misled by all this into thinking that toxic tort suits are therefore an overwhelmingly popular legal option for those seeking to protect the environment. They are not, in large part because toxic tort cases are in fact very difficult to win. This is another example of the relevance of the point we made a number of times in Chapter 2 (and will continue to make in this and the following chapters): the legal government of the environment (and climate change) is never easy, never straightforward, and never complete or perfect; it cannot expect to ever finally escape from the long record of serious failures which accompanies any branch of legal-political government wherever it operates for a long period of time.

The primary difficulty for toxic tort litigation is that the plaintiff has to be able to identify through expert scientific evidence both the nature of the toxin claimed to be involved and the causal link between the plaintiff's exposure to this toxin and the injuries suffered. Part of the predicament here is that the tort principles of the nineteenth century, which remain important to the operation of tort cases in the twenty-first century, were premised on the assumption that there is usually only a single line of causation linking the defendant's negligence and the plaintiff's injury. Yet in the course of the twentieth century and into the twenty-first, science and medicine have developed to the point where an injury is now unlikely to be attributed by scientific or medical experts to a single cause. Furthermore, the medical model employed to determine the actual cause of any injuries is no longer based on direct causality. Rather, it is based on a 'multi-factor approach' which understands 'the problem of health in a broader context', whereby a disease can be attributed to exposure to a variety of toxins as well as to the lifestyle and patterns of consumption of potential plaintiffs (Lanthier and Olivier 1999: 75). An example will help us to establish this point. It features a series of events in the Australian city of Brisbane in the early years of the twenty-first century.

Ten women who worked in the Brisbane television news studios and offices of the Australian Broadcasting Corporation (ABC) all developed invasive breast cancer within a relatively short time (Swan 2007). The unhappy coincidence of ten women in the same workplace developing breast cancer in quick succession is, on the face of it, disturbingly insidious. Yet initial investigations were not able to identify a cause or to offer a satisfactory explanation. A more thorough inquiry, led by Professor Bruce Armstrong, seemed at first to be more promising. But subsequently, despite finding that the incidence of breast cancer in women working at the studios was ten times the rate of that of women of the same age in the general population (Armstrong 2006), it too was not able to identify a specific cause, though it did conclude that the

Brisbane ABC studios presented an unequivocal risk to health. The studios have now been abandoned and all ten cases of the cancer have been filed as a rare 'cancer cluster' (Swan 2007). At the time of writing, no toxic tort suit has been launched in relation to this series of cancers.

As we suggested above, scientists, including medical scientists, cannot always identify the 'guilty' toxin, even when the effects of exposure to a toxin are evident, and cannot always track the actual effects of known toxins, even when the effects of exposure to these toxins, such as lead, are well documented (Cranor and Eastmond 2001: 11).

To compound the difficulty on this issue, if a causal link is established in any particular case, a court may still determine (as we will illustrate later in the chapter) that, at the time of the exposure, knowledge of the particular toxic hazard was so limited that the harm the plaintiff has suffered was not 'foreseeable' by the defendant. In such a case the court would likely rule that the defendant could not have known of the risk and so could not therefore have been expected to limit the plaintiff's exposure. As such, the defendant would not be found to have breached the duty of care owed to the plaintiff and would not be judged legally liable.

This type of adjudication of the defendant's duty of care and of the unlikelihood of the defendant foreseeing harms, as well as the difficulty of amassing sufficient evidence of causation, makes toxic tort litigation so difficult that it can pile more anxiety for any would-be plaintiffs on top of the anxiety produced by their injuries. We will offer further discussion of this type of 'risk anxiety' later.

A consequence of the many obstacles being discussed here is that the outcomes of toxic tort litigation have come to play a not insubstantial role, across the populations of modern law-and-politics countries, in setting standards for the toleration of toxic risk (Havemann 2003: 38; Rabin 2001: 352–53). In this way, toxic tort often comes to serve as a vehicle for wider articulations of the relations between individual bodies, hazards, and the environment generally (Kroll-Smith and Westervelt 2004; Lee 2000; Rabin 2001). These points are illustrated by the example of the ABC Brisbane breast-cancer cluster, which was widely reported and discussed even though it was not litigated. It was as if the difficulty of litigation, in what was seen by the majority of those members of the public who aired their views as an instance of an event which 'deserved' legal redress, served as a spur to claims that 'something' must be done to limit such environmental harms. This was also the case with another recent example.

This second example took place in Esperance, a medium-sized town on the southern coast of Western Australia. Famous for its pristine beaches and unspoilt coastline, the town was shocked in 2007 when over a thousand native birds died suddenly and rainwater tanks, vegetable patches, and backyard chicken runs were found to be toxic. The blood samples taken from six children were also found to contain lead at levels above recommended international health guidelines (*The West Australian* 2007: 8). In this case, however, unlike the Brisbane case, the source of the toxin was readily identifiable. In short, parts of the town were poisoned by lead carbonate, which at that stage was being carried through the town in uncovered rail wagons and loaded onto ships at the town's port.

Yet even here toxic tort litigation was not launched, largely because of the difficulties described above. As one might imagine, this incident produced at least as much risk anxiety as did the Brisbane case. Both incidents, this is to say, tapped into an anxiety that the environment in which we live and work almost certainly contains toxins which we know nothing about or, worse perhaps, know something, but too little to do anything to limit the harms they might produce. In a radio interview conducted by Dr Norman Swan, the presenter of the ABC's 'Health Report', one of the ABC journalists in Brisbane who had developed breast cancer, Nadia Farha, articulated this anxiety very clearly. Upon learning of the findings of the Armstrong Inquiry, she said:

> I was really upset, I think that's where it actually hit me that maybe working for the ABC at that particular site I could have got my breast cancer. And you think that you'd worked for an organisation all of these years, and you've given them the best of your career to find that working there may eventually kill you brought me down to earth pretty much in a hurry. I was really shocked, and I think some of the women weren't that shocked because I think that's what they expected but I hadn't expected that.
>
> (Swan 2007, quoting Nadia Farha)

The fear, anger, and frustration that Nadia Farha articulates here might be considered a perfect real-life example of Douglas and Wildavsky's astute observation (1982: 10) that when it comes to the environment, people are afraid of 'nothing much ... except the food they eat, the water they drink, the air they breathe, the land they live on, and the energy they use'.

In other words, risk anxiety, which is born partly of the toxic exposures themselves and partly of the difficulties involved in using the common law to address environmental hazards, is largely about the strong need of the people involved in particular cases to identify the source of their toxic exposure, and, moreover, to hold somebody accountable. In the two examples discussed so far, the risk anxiety was heightened by the very 'normal' circumstances in which the toxins entered the people's lives. They were exposed to toxins not while working in a dangerous mine or in a nuclear facility, but while they were going about their routine activities in places they had no reason to suspect were potentially poisonous.

While no toxic tort suit has been so far opened to address the issues in the two examples under discussion, the issues involved have not by any means been forgotten. Various reports and documentaries have led to widespread calls concerning: the need of the people exposed to such dangerous toxins to be heard (particularly the women with breast cancer in the Brisbane example); the need for the matters to be more thoroughly investigated; the anxiety those directly affected and others near them continue to endure about the toxicity of their immediate environments; and their unresolved anger at becoming the victims of toxic hazards (*ABC Online* 2007a; 2007b; Swan 2007; *The West Australian* 2007: 8). These sorts of concerns may be less tangible than diagnoses of cancer or increased lead levels in blood samples, but they are nevertheless an important part of the challenge faced by law and politics generally, as they seek to govern the environment, and of the challenges faced by common-law toxic tort in particular as it seeks to address aspects of the environment. We will return to the particular challenges of toxic tort after we say something more substantial about the notions of risk and risk anxiety.

The centrality of the notions of risk and risk anxiety in toxic tort

Inasmuch as actions in toxic tort endorse and articulate risk anxiety, particularly in instances where the litigation gives publicity to latent hazards that had previously been little known (Rabin 1993: 126), toxic tort operates in a manner that is consistent with the influential observations about risk made by the social theorist Ulrich Beck in building his argument for the emergence of a 'risk society' in the modern world. For example:

> Dangers, it would seem, do not exist 'in themselves', independently of our perceptions. They become a political issue only when people are generally aware of them; they are social constructs which are strategically defined, covered up or dramatized in the public sphere with the help of scientific material supplied for the purpose.
>
> (Beck 1999: 22)

In the light of Beck's observations, we might ask some pointed questions about our two main examples. If either or both of the toxic exposures at the ABC Brisbane studios and the one in Esperance were to become the subject of toxic tort action, and if we could think past the obstacles to litigation in these two cases, what would be the relevance to any such litigation of the anxiety and concerns of the potential plaintiffs? How, if at all, would their quite subjective and particular appreciations of the toxic hazard to which they have been exposed be assessed? Does a plaintiff's risk anxiety have any legitimate role in the determination of a toxic tort claim?

Although toxic tort litigation clearly does not offer the same forum for exploring these matters as does a television documentary or a news report, it is not utterly divorced from their significance. As former High Court of Australia Justice Michael McHugh observes, the applications of tort doctrine 'depend not only upon the ascertainment of facts but on a moral and social evaluation of those facts' (McHugh 1989: 13). (In this instance, it is important to note, Justice McHugh is not using the word 'moral' in its metaphysical sense, to evoke some supposedly timeless and universal source of morality; instead he is using it in its common-law sense, where it marks a domain of individual responsibility.)

In writing about the relations between law and risk, Jenny Steele (2004: 9) argues that risk calculation, risk vocabulary, and risk management techniques are taken up in litigation in specialized ways, in a bid to facilitate appropriate judgments regarding the limits and responsibilities associated with risk taking, although in other arenas they can serve quite a different purpose. By this she means that 'risk' has many meanings and functions. On the one hand, risk entails a quantitative, actuarial calculation which measures the probability of an event occurring, and as such it is a commodity that can be traded (whether through a stock market or at a race track). On the other hand, it is readily employed in everyday speech and mediates the relations people have with certain activities (Lupton 1999). 'Risk', this is to say, is a label which attaches to hazard, danger, and uncertainty in both a scientific and an everyday sense (Ewald 1991). None of these understandings of risk is isolated from the others, and they are often joined together in particular contexts, such as the courtroom, to create new ways of understanding risk suited to particular tasks, including assigning legal liability (Valverde et al. 2005: 86).

In the context of our discussion, then, risk is a way of conceptualizing, measuring, and even just discussing environmental hazards. It is not only a label which attaches to some hazards. In line with this, assessing the role of risk in the toxic tort litigation process is not always straightforward. Steele points out (2004: 9) that the courts necessarily engage in a quite specialized legal mode of addressing risk, with specialized terms and techniques used to determine the appropriate limits and responsibilities of risk taking in any particular case. By Pat O'Malley's (2004) account, whenever the law confronts risk taking it does three things: law assigns a moral value to the risks it is assessing (again, this is 'moral' in the legal sense – individual responsibility); law either fully accepts the need for these risks or sees a reason to curtail them; and law seeks to determine the limits of legal liability of the risk taker(s).

These points further establish the argument that by itself an exposure to a risk is not enough to found a tort claim. Plaintiffs must be able to demonstrate that they have actually suffered an injury attributable to the defendant's negligence. Even though toxic tort may become an occasion for the litigious expression of popular risk anxiety, the traditions of common-law litigation require each party to support its case with expert testimony as to the actual, rather than simply apprehended, nature of the risk to which the plaintiff was exposed. Scientific calculations and assessments of the harm plaintiffs might have suffered and the actual risks to which they were exposed are central to the resolution of any toxic tort claim, as we will illustrate in a later sub-section. Plaintiffs must provide medical evidence of their actual injuries, but that alone will not be sufficient. Whether the defendant should be held responsible for the plaintiff's injury will depend in large part upon the risk calculations of experts such as the epidemiologist and the environmental engineer. The application by a court of various technical or scientific risk technologies thus allows it to make an assessment of (a) whether the defendant should have foreseen the possibility that the plaintiff would suffer the injury and (b) the causal nexus between the injury and the toxic exposure or environment.

More than this, actions in tort take place against and within the operation of various insurance schemes, whether state operated or privately operated. Noted commentators on the development of tort law have identified a number of key effects of insurance and insurability upon the tort system and the conduct of litigation. Perhaps the most profound effect is that identified by John Fleming, who argues (1992: 11) that without liability insurance 'the tort system would have long ago collapsed under the weight of the demands placed upon it'. By this argument, when liability insurance became more commonly used across the course of the twentieth century, the historical conservatism that had previously militated against the emergence of new forms of action was displaced. As Jane Stapelton puts it (1995: 820), the increasing 'public policy reliance' on a prudential response to risk has been paralleled by 'a general broadening of the catchment of situations recognised by the courts as giving rise to tort entitlements'.

It might not be saying too much to say that liability insurance has transformed the very nature of tort law. Morton Horwitz, discussing the impact of liability insurance on the development of tort law in the United States, goes some way down this road:

> The individualistic world of Warton's 'moral causation' [individual responsibility] and 'free agency' had begun to be transformed into a world of liability insurance in which the 'legislative' question of who should pay would ultimately undermine the self contained, individualistic categories of private law.
>
> (Horwitz 1982: 211, quoting Warton)

But Joanne Conaghan and Wade Mansell (1993: 11) are slightly more cautious: 'individual responsibility continues to be the perspective that informs most cases, collective responsibility and loss spreading through the mechanism of liability insurance tends to be the incidental by-product'.

A different perspective is offered by François Ewald, who, in his genealogical study of insurance technologies (1991), explores the ways in which forms of rationality employed by different types of insurance directly challenge the juridical practice of assigning responsibility according to legal right (see especially 1991: 201). In other words, where both Horwitz and Conaghan and Mansell put more emphasis on one side or the other of the law–insurance nexus, Ewald proposes that law and insurance actually employ different modes of assigning responsibility and different modes of compensating for loss. In offering this proposition, Ewald argues that the rationality of the law and the rationality of insurance may be applied to the same object to different ends: 'Insurance and law are two practices of responsibility which operate quite heterogeneous categories, regimes, economies; as such, they are mutually exclusive in their claims to totality' (Ewald 1991: 201). The fundamental difference between law and insurance, Ewald argues (1991: 203), is that where the juridical focus on the occurrence of an event is singular and moral (again in the sense of individual responsibility), insurance eschews any question of morality and instead factors the probability of any event occurring and/or recurring in a predictable fashion.

Treating risk in terms of an actuarial calculation or in terms of scientific analysis is not, we would suggest, consistent with the most popular understanding of risk, which is about potential dangers and whether one wishes to confront them (even enjoy them) or avoid them. In this everyday sense, risk assessments are not statistical calculations or scientific judgments, they are the products of humans interacting with other humans, interactions which, over centuries, or even millennia, have spawned many codes – of religion, morality (in both its strict legal sense of individual responsibility and its metaphysical sense of universal, timeless sets of criteria for judging the operation of law and politics), propriety, politics, law, government, organization, commerce, and so on. This complex of understandings and codes of risk is not limited by the confines of scientific discourse; though the complex does not shun science, it actually welcomes scientific insights and evidence, albeit usually in anecdotal form (Douglas 1992: 24). As Jacqueline Peel puts it:

> 'lay' people evaluate health and environmental threats according to a different set of criteria than may be reflected in expert assessments. ... [T]he risk perception of lay members of the community appears to be influenced by various contextual factors that lie outside the realm of scientific research.
>
> (Peel 2005: 68)

The everyday or lay sense of risk carries with it the expectation that individuals will monitor themselves and will be risk averse much more often than they will be risk takers. Because of this, argues Mary Douglas (1992: 16), in everyday settings risk has become inextricably linked with blame. Whereas in courts the technical calculations of epidemiological risk are treated as crucial objective measures, in the everyday world these calculations are usually shunned, in favour of ascribing culpability along the lines of some supposedly universal, timeless moral code. But, as we have already glimpsed several times, this does not mean that courts deal exclusively in hyper-rational calculations of risk, such as those produced by scientists or actuaries. The toxic tort courtroom appears to rely as much upon everyday assessments of the parties' conduct (with all of the history such assessments carry with them), as they do on scientific argument and testimony, actuarial calculations, or finer points of law.[3]

Risk, all this is to say, is important to the operation of toxic tort, yet, because of its historical richness, it is a source of the difficulty faced by those who wish

---

3 Robert Rabin (1993: 122), for example, describes tobacco litigation in the United States as 'a last vestige of a vision of nineteenth century tort law as an interpersonal morality play'.

to pursue toxic tort suits (which is yet more evidence for our argument that the legal-political government of the environment (and climate change) is never easy, never straightforward, and never complete or perfect). This brings us back to the actual operation of toxic tort cases (we will discuss risk again in a later sub-section concerned with the way courts deal with it).

Tort goes to court

In discussing risk and risk anxiety we speculated about just how the potential plaintiffs in both the Brisbane and Esperance examples might approach toxic tort litigation were they to pursue this difficult course of action. We can now add to that hypothetical discussion by asking just what any potential legal team might be looking for in deciding whether a suit brought on behalf of those affected by these two incidents could be successful. To begin, the lawyers involved would be forced to admit that the source of the toxicity in the ABC example is elusive, while in the Esperance example it is pretty clear cut. So they would almost certainly conclude, quickly if reluctantly, that it would not be sensible for the ABC women to sue, simply because, as was discussed earlier, establishing causation and responsibility for the exposure in that instance appears to be beyond current science and so beyond current law. But the Esperance example looks at first glance to be a different matter.

The circumstances by which some people in Esperance unwittingly came to be vulnerable to lead poisoning, simply by growing their own vegetables, collecting rainwater in tanks, and so on, makes them into more likely plaintiffs. Indeed, the firm we mentioned earlier, Slater and Gordon, visited Esperance and undertook their own investigations into the circumstances of the toxic exposure. But because the only publicly reported action the firm took was to send a letter of demand to the relevant Western Australian Government Minister, urging him to establish a compensation scheme which would allow those affected to avoid litigation (*The West Australian* 2007: 8), we have to conclude that even in this example, and even with this law firm, a successful toxic tort case would be a bridge too far. Slater and Gordon seems to have determined that because toxic tort litigation is all too often an inefficient and compromised endeavour, the letter of demand was the only way to proceed.

In other words, the first thing one should know about the operation of toxic tort as a particular form of the legal-political government of the environment is a point that keeps forcing its way into our considerations: that the bar in toxic tort is set very high, mainly because the risks of modern society are difficult to identify with certainty and the causal nexus between injury or damage, on the one hand, and risk or hazard, on the other, is difficult to establish (Cannon 1998; Lee 2000). Perhaps the second thing one should know about the actual operation of this form of law concerns another matter touched on earlier, the fact that nearly all corporations and government agencies would wish to use their insurance cover to circumvent any matter reaching court. But this 'second thing one should know' is the subject of some debate.

Peter Cane argues (1997: 13) that because the most significant characteristic of any tort action is the correlativity of the parties – 'Every cause of
action in tort has elements concerned with the conduct of interacting parties'
– the real implications of insurance coverage generally do not figure in the
courtroom. This is because, he says, 'it will always be possible to rationalize
a rule of tort law in terms of principles of personal responsibility ... even if
it also rationalizes the decision in terms of loss spreading' (Cane 1997: 230).
Against this, Horwitz, arguing that the correlative character of tort litigation
described by Cane is to some extent a chimera, says of the usual tort case:
'Liability for injury has become just another cost of doing business, which
could be estimated, insured against, and ultimately included in the price paid
by the public' (Horwitz 1982: 211).

When it comes to standard personal injury claims, we think Horwitz's assessment of the current trajectory of tort litigation is the more accurate, which
means that one must take the actions and likely actions of insurers into account
in trying to understand the operation of toxic tort cases. Fleming (1992: 10), for
example, observes that a defendant's insurance may undermine any deterring or
punishing effect that might have followed a finding of legal liability. Indeed,
insurers may have the capacity to dominate the litigation process, though this
is not to say they participate in this process as if factors beyond their own actuarial calculations are irrelevant. Rather, it is to say that, because determination
of legal liability in toxic tort cases inherently incorporates an assessment of
the culpability of not only the defendant but also the plaintiff, insurers have
an interest in moving beyond their actuarial domain at some stages of a case.
The culpability of plaintiffs is bound up with assessments of whether they were
suitably risk averse, in a non-actuarial sense (that is, in the complex everyday
sense we outlined above). In making such assessments the courts weigh up
competing accounts of the type of risk to which any plaintiff was exposed.
These competing accounts come not just from the experts who provide technical assessments of the hazard or harm to which the plaintiff was exposed; they
come also from the parties themselves, and from other lay witnesses, such as
workmates, whose role is to provide the court with a thorough appreciation of
the environment in which the plaintiff may have been exposed to risk.

As we have noted a number of times, a central feature of all toxic court
cases is the role of expert evidence in determining causation. A good deal of
scholarly discussion of toxic tort cases focuses on the inevitable disputes over
the veracity of scientific evidence (see, for example, Cranor and Eastmond
2001; Edmond and Mercer 2002, 2004; Huber 1991). We acknowledge the
importance of this literature, but our main concern is with the complex relations between scientific evidence and the evidence that is provided by non-
expert plaintiffs and by other non-scientific, non-legal witnesses who have
themselves experienced the toxic exposure (in this we are closer to scholars,
such as Jasanoff 1995; Morrow 2000; Rabin 1993, 2001; Toffolon-Weiss and
Roberts 2004). While courts are always likely to consider scientific evidence
to be objective evidence, it must always be remembered that non-scientific,

often quite subjective evidence about risk will prove no less important, for both the plaintiff and the defendant. We will shortly provide more detail on the role of these different types of evidence in court, after our promised brief discussion of the way risk is treated in court.

Risk goes to court

To sum up the types of evidence likely to figure in any toxic tort case, it is helpful to divide risk into four types of calculation (three of the types have been introduced already, the fourth has not; it will be the main topic of this sub-section). The four types are: actuarial calculations about risk, scientific calculations about risk, everyday calculations about risk, and clinical calculations about risk. In an important sense actuarial calculations about risk, as they operate in the insurance industry, have facilitated the emergence of toxic tort as a specific form of negligence action (Fleming 1992: 15). The scientific calculations that are made about risk in toxic tort cases by, for example, epidemiologists, have become the main means of marking the boundary between tenable and untenable claims (Edmond and Mercer 2002; Jasanoff 1995: 16). Everyday calculations about risk, unlike the other forms of calculation being considered here, have been the subject of very little academic debate. Nevertheless, the particulars of any plaintiff's claim will be scrutinized in a case through the lens of just such everyday calculations, by both the plaintiff and defendant (Valverde et al. 2005: 86; *Seltsam v McGuiness* 2000). Clinical calculations about risk aim to combine the scientific and everyday calculation into a new entity. Where epidemiological assessments of risk involve scientific calculations of factors across populations, the clinical risk approach involves scientific calculations of factors specific to the domain of everyday risk, that is, calculations made in terms of specific individuals. For the clinical way of calculating risk, then, each 'individual is treated not simply as representative of a risk category but as a unique case to which certain risk factors apply' (O'Malley 2004: 25). Courts hearing toxic tort cases have become more and more interested in clinical assessments. This is because the court's own assessment of a plaintiff's case not only involves situating the incidence of toxic exposure in the context of accepted knowledge about the risk associated with exposure to that toxin; it also involves an appreciation of the parties' personal experience of risk, and an appreciation of their behaviour in the face of such exposure.

This form of clinical calculation about risk thus entails at least a touch of the notion of 'moral causation' (individual responsibility), which we have met on a number of occasions:

> Under the regime of juridical responsibility ... The accident is due to some individual fault, imprudence or negligence; it cannot be a rule. Moral thought [in the legal sense] uses accident as a principle of distinction ... a unique affair between individual protagonists.
>
> (Ewald 1991: 203)

Science goes to court

Despite the rising importance of clinical calculations in toxic tort courts, especially because of their capacity to promote certain forms of the everyday calculation of risk, and despite the ongoing importance of actuarial thinking, we suggest that expert scientific evidence is ultimately the most important factor in courts' determinations of causation. This is to say that when it comes to a court determining the risk presented by exposure to a particular toxin or toxic environment, science is king. Inevitably, the manner in which scientific disciplines such as epidemiology (the discipline on which we focus most of our energies here) conceive of and articulate risk has the greatest impact upon toxic tort decisions. Evidence from epidemiology – the study of 'the incidence, distribution and aetiology of disease in human populations' (Freckelton 2000: 133) – seems to be almost irresistible for toxic tort courts.

In the conduct of a particular case, the strength of the associations the epidemiologist might find between a possible cause and the incidence of a disease will depend upon excluding alternative causes as well as upon the size of the population or cohort under investigation (Billauer et al. 1989: 66). While reliable epidemiological studies are available for only half of the known and likely human carcinogens (Cranor and Eastmond 2001: 10), their role in toxic tort litigation has not been diminished by this fact, not only because such studies allow the court to make some sort of objective assessment on the question of causation, but also because they provide a reliable device to help courts determine whether a defendant should have been able to foresee a particular toxic risk. In this way, epidemiological expertise forms part of the 'evidence of possibility' that courts use to map the potential hazards to which a plaintiff may have been exposed, potentially on a global scale. The court can then use this map to assess the conduct of all parties. As Edward Christie notes:

> A feature of a toxic tort dispute is that a plaintiff can rarely introduce particularistic evidence which directly addresses the question of proof of causation in the individual plaintiff's case ... [P]arties must, instead, rely on evidence which indicates an increased risk, or increased probability, of disease incidence following exposure to a specific chemical.
>
> (Christie 1992: 303)

Somewhat paradoxically, epidemiology has become a dominant form of scientific evidence in toxic tort cases in spite of the fact that its calculations of risk are generally not created specifically for legal purposes. Because of this, the value and/or methodological reliability of epidemiological evidence is regularly contested in court (Billauer et al.1989: 66). It has to be added, however, that such contests are in one sense simply an expression of the fact that there is no such thing as complete scientific certainty in any domain, especially in court. As Peel observes (2005: 35), 'various types of logical errors' are inherent in all scientific knowledge claims. One should never lose sight of the fact that science in the courtroom is employed in an adversarial context, which is different from the context in which it is employed in the laboratory, and hence the testimony of expert scientific witnesses is 'strategically framed' for the court's consumption (Jasanoff 1995: 48). Indeed, the Australian courts freely acknowledge:

> The pragmatic assessment of probable cause as a basis for tortious liability cannot be wholly constrained by the scientific and philosophical purity of epidemiology, which essentially depends upon a comparison of the data obtained in controlled circumstances.
> (*E.M. Baldwin & Son Pty Ltd v Plane & Anor; Jsekarb Pty Ltd* 1999)

Steele argues (2004: 9) that courts pragmatically employ a decision-making model which 'constructs a moment of decision which may be purely hypothetical and uses this to draw the "right" conclusion'. This is a dynamic apparent in the leading Australian case of *Seltsam v McGuiness* (2000) (hereinafter *Seltsam*), in which the New South Wales Court of Appeal held that epidemiological studies 'should be regarded as circumstantial evidence which may, alone or in combination with other evidence, establish causation in a specific case'. The Court recognized epidemiology's 'potential utility' in toxic tort cases on the grounds that it may be able to 'fill the gap' whenever 'medical science cannot determine the existence or non-existence of a causal relationship for purposes of attributing legal responsibility' (*Seltsam* 2000: 277). However, despite Chief Justice Spigelman's thorough examination of the utility of epidemiological evidence in *Seltsam*, in his judgment he does not identify the point at which epidemiological or other forms of scientific evidence should be considered by the court to be compelling (Freckelton 2000: 140). This is the case largely because, as we have argued previously, courts always determine that question in the context of a wider body of non-scientific evidence, which combines, in Spigelman's words, with the epidemiological evidence like 'strands in the cable', and from which causation might be inferred by the court as a matter of common sense (*Seltsam* 2000: 278). Of course this leaves open the question of whose common sense is to be used. Is a judge's common sense in any way akin to that of the journalist working in the ABC's Brisbane Studios, whose own common sense tells her that despite the lack of a conclusive scientific explanation, coincidence alone does not explain why ten women in the one workplace have all developed invasive breast cancer (Swan 2007)? Jasanoff observes (1995: 59) that the exercise of the court's discretion (its common sense) is inevitably shaped by the degree to which 'judges are swayed by their perceptions of what "science" is and who is a "scientist" when they certify an expert's credibility'.

We turn now to some American cases. Debate about the proper interpretation of the United States Supreme Court's ruling on the standards that should apply in determining the admissibility of scientific expert evidence in *Daubert v. Merrell Dow Pharmaceuticals, Inc.* (1993) (hereinafter *Daubert*) demonstrates how assessments of risk such as those produced by epidemiologists are not neutral elements in the legal assessment of risk (this case involved a suit against the manufacturer of the anti-nausea drug Bendectin, which had been prescribed for pregnant women to alleviate morning sickness and was later associated with birth defects in the babies of many of those who had taken it). The debate regarding the proper interpretation of *Daubert* is about more than simply what type of science should be recognized in the courtroom. For Jasanoff (1995: 65–66), this case involves 'the social and moral viability of particular technological choices'.

The Supreme Court took the opportunity in *Daubert* to develop criteria which would allow courts to secure 'relevant and reliable evidence' and overcome what it saw as the inconsistent jurisprudence that had developed in relation to the admissibility of expert evidence (Edmond and Mercer 2004: 234). When determining the validity and probative value of scientific evidence, such as the 'falsifiability' of scientific criteria relied on in expert testimony, the Court was persuaded to privilege scientific criteria by a number of influential *amicus* briefs from corporate-sponsored think tanks (Edmond and Mercer 2004: 244). The *Daubert* decision listed particular criteria, which it hoped trial judges would take into account in assessing claimed expertise in future cases, though the Court indicated that these criteria were intended only as a guide.[4]

However, as is often the case when a superior court attempts to set criteria 'only' for guidance, the *Daubert* criteria soon lost their intended flexibility and through their very applications were transformed into a checklist for the application of a fixed rule. Gary Edmond and David Mercer note (2004: 244, 250) that in subsequent decisions *Daubert* has become the basis for imposing a much more stringent admissibility threshold. For their part, Carl Cranor and David Eastmond suggest (2001: 9) that courts have 'misunderstood or learned the wrong lessons' from *Daubert*. They argue (2001: 6) that the *Daubert* reforms have thus locked out plaintiffs who have based their claim on 'reliable, but not ideal, scientific evidence'. In a similar vein, Edmond and Mercer cite the fate of *Newman v Motorola Inc.* (2002) (hereinafter *Newman*) in support of the argument that, '*Daubert*-inspired quests to establish scientific truth … may assist in discouraging ongoing legal scrutiny of intransigent

---

4 The Court listed three criteria that, though used in scientific circles, were new to courts. By the first of these, the Court sought to introduce Popper's proposition about the necessity for any scientific theory or technique to be falsifiable (Edmond and Mercer argue (2004: 236) that some of the Court's handling of this proposition is 'inconsistent with some of Popper's aspirations for falsification'). The second criterion focused on whether an explication of any scientific theory being used in a case had been published via a peer review process (here the Court acknowledged arguments that peer review and publication do not guarantee reliability). By the third criterion, the Court proposed that at least some consideration be given to any known error rate of any scientific technique involved in evidence before a court (Edmond and Mercer 2004: 234).

scientific controversies involving uncertain risks' (Edmond and Mercer 2004: 243). Because *Newman* was one of the first cases relating to the hazards of cell phone use, its failure has been significant for a large cluster of potential toxic tort claims relating to the harm caused by these types of phones (known as mobile phones in other parts of the world) (Edmond and Mercer 2004: 239–40; Capriotti 2002).

Justice Blake, in a decision that was affirmed on appeal (*Newman v Motorola Inc* 2003), rejected the evidence of the plaintiff's expert epidemiological and oncological witness, Professor Lennart Hardell. The judge's thinking, Edmond and Mercer argue (2004: 240), was a 'strategic manipulation of the *Daubert* criteria'. To some extent, however, one can understand Justice Blake's ruling, inasmuch as Hardell's research and opinion was not consistent with the bulk of the then current science on the biological effects of electromagnetic radiation and cell phone use. This is to say that the scientific debate about cell phone use was not settled at the time and has not been settled since. There simply has not been sufficient scientific research to produce a large enough cohort of epidemiological studies of the effects of electromagnetic radiation generated by cell phones (Capriotti 2002: 4). Nonetheless, public concern about the risk of using cell phones, coupled with the want of scientific certainty on the issue, has been enough to generate independent inquiries in several Western countries (Capriotti 2002: 2–3; Peel 2005: 108–09). As Cranor and Eastmond note (2001: 13), even when a potential toxic hazard comes under scientific scrutiny, it takes considerable time and resources to accumulate a body of reliable scientific evidence evaluating its effect on humans and the environment and 'it takes longer still to establish a scientific consensus'.

In the case of *Newman*, Professor Hardell's expert testimony was rejected because Justice Blake disapproved of his methodological approach, leading her to conclude that his research failed to deliver 'exactness and certainty'. Edmond and Mercer argue (2004: 241–42) that her approach to the admissibility of Hardell's evidence demonstrates how 'ideal images of the scientific method can be used in legal settings to help deconstruct or marginalize particular forms of expertise', and in so doing 'restrict the entry of (novel) scientific claims'.

In Australia, by comparison, the current 'commonsense' approach to causation adopted by the courts favours a broader, less idealized reliance on expert scientific evidence. In this way, the Australian courts recognize that scientific evidence is unlikely to fully determine questions of legal causation. But, as we emphasized earlier, what counts as common sense varies. Were an Australian judge to consider common sense to mean total commitment to scientific evidence, the *Daubert* thinking might gain a foothold in Australian jurisdictions. A greater emphasis on scientific evidence might then see the downgrading in Australia of the plaintiff's lived experience of the hazard in favour of what Karen Morrow, describing some British and Irish nuisance cases, calls a 'harder' approach to causation (Morrow 2000: 144, n. 3). We doubt this is going to happen in Australia, but we think it worth considering some of Morrow's arguments about 'hard' and 'soft' approaches by courts to scientific evidence.

In one of the nuisance cases she deals with, *Graham and Graham v Re-Chem* (1996) (hereinafter *Graham*), Morrow describes how the court was not satisfied that the plaintiff's evidence established a causal nexus between the damage suffered by the plaintiff and the emissions from the defendant's incinerator. Their evidence was judged '"very confused and confusing, contradictory and riddled with inconsistencies"' (Morrow 2000: 148, quoting *Graham*). The fact that there had been findings by official inquiries regarding the hazards of the type of incinerator used by the defendant was also considered to be of little consequence, even when coupled with the plaintiff's testimony. The reports documenting the risk from this type of incinerator were seen by the court to be too general and so were not treated as evidence for the claim that actions of the defendant specifically caused the plaintiff's injuries. In essence, the plaintiff's case failed, in the court's view, because the plaintiff had not provided '"detailed clinical, pathological and histological evidence of ... toxic insult"' (Morrow 2000: 148, quoting *Graham*).

Inasmuch as a court has immense power, simply through language, to characterize a plaintiff's 'commonsense' evidence as either more or less authentic than the scientific evidence presented, it is possible to point to cases in which a 'hard' approach and a 'soft' approach confront one another, as Morrow has done in the case of *Hanrahan v Merck, Sharp & Dohme (Ireland) Ltd* (1988) (hereinafter *Hanrahan*), in which the original judgment was appealed. Where the original judgment was not dissimilar to *Graham* in its 'hard' approach, on appeal Justice Henchy found that the defendant's expert scientific evidence, based on readings of emissions from the plant and computer models, did not pay due attention to the 'real physical context of the emissions' as evidenced by the plaintiff's witnesses. He held that, '"Theoretical or inductive evidence cannot be allowed to displace proven facts ... it would be to allow scientific theory to dethrone fact"' (Morrow 2000: 146, quoting *Hanrahan*).

The degree to which a court aligns 'legal causation with scientific causation', then, will not only determine what type of science will count in the litigation, it will also have an impact on the significance attached to other forms of non-scientific evidence (Edmond and Mercer 2002: 103). Robert Lee suggests (2000: 86) that because 'the public experience of risk is not one of unthinking acceptance of a position expounded by experts, nor is it a simple choice between expert positions', we should not be surprised that non-scientific or lay evidence is often very important in court. Notwithstanding the trend in some post-*Daubert* litigation in the United States and the other 'hard' approaches Morrow highlights, most toxic tort courts in the common-law world have not taken up the habit of calculating risk solely according to expert scientific evidence. In *Chappel v Hart* (1998), for instance, the Australian High Court observed that causation is 'a question of fact resolved as matter of commonsense and experience, the conclusion is often reached intuitively' (*Chappel v Hart* 1998: 256).

In this way, a plaintiff's subjective appreciation of toxic exposure and its consequences remains no less important as evidence to the court's determination of legal liability than does the defendant's conduct. This dynamic is readily apparent in Justice Seaman's decision in the Western Australian Supreme Court case of *Napolitano v CSR Ltd & Anor* (1994) (hereinafter *Napolitano*). The plaintiff in this case, Mr Napolitano, had worked for two years for the defendant in its blue asbestos mine in Wittenoom, Western Australia. He sued the defendant in tort not only because he had developed mesothelioma (a fatal asbestos-related disease), but also because he had developed a psychiatric illness, which he attributed to his longstanding fear that he would succumb to mesothelioma. The defendant admitted liability for the plaintiff's mesothelioma, but denied liability for his psychiatric illness. The Court regarded the 'question of liability for psychiatric injury to be a matter of major complexity' (*Napolitano* 1994: 6). However, the complexity had to be dealt with in an expedited hearing – occasioned by the knowledge that the plaintiff was likely to die imminently – and the judgment is characterized by a pragmatic and candid reading of the evidence, with little of the rationalization that usually attaches to judicial decision making. Because of this, Justice Seaman has come in for criticism from at least one legal scholar, who noted the judgment's 'lack of analysis of principle or authority' (Mullany 1997: 137). For our book, by contrast, the judgment is all the more instructive for the fact that the judge, in ruling that the defendant was liable for the psychiatric illness, demonstrated the importance to the Court of a clinical gaze on the plaintiff's personal experience of living with the risk of such massive exposure to asbestos.

To put it another way, in deciding, on the balance of probabilities, that the plaintiff's 'illness of mind' was caused by the defendant's conduct, the Court put a premium on 'applying common-sense to the facts' (*Napolitano* 1994: 20). Justice Seaman clearly considered and rejected the legal precedent that fear of contracting a disease is not on its own a basis for any form of compensation beyond payment for medical monitoring of the plaintiff's health: 'It seems to me the controlling features of liability for psychiatric illness cannot be determined in the abstract and this case cannot turn on statements of policy apt to very different factual situations' (*Napolitano* 1994: 22). In the Court's opinion, then, the particulars of Mr Napolitano's experience of fear, made worse by the psychiatric diagnosis (and ultimately vindicated by the fact that he died of the very disease he feared), was more important than the mere fact that he suffered from a recognizable mental illness. In the judge's words, it was Mr Napolitano's 'very long period of misery and anxiety and fear', compounded by 'the distressing phenomena of asbestos-induced illness in his fellow Wittenoom workers who were his friends and in particular Mr Cinquina [his best friend]', that led to the 'depressive illness' (*Napolitano* 1994: 20–25). As Nicholas Mullany describes this judgment (1997: 137),

'One gleans the impression from his judgment that Seaman J simply formed the opinion that it was just in the particular circumstances to compensate the plaintiff'.

Conclusion

In this chapter we have added another dimension to our answer to the book's key question – how do law and politics, entwined in various complex relations with sovereignty, the state, morality, religion, economy, aesthetics, and science, attempt to govern the environment and climate change? This new dimension is law *in situ*, this time in its micro common-law form, using the example of toxic tort. After insisting that under legal-political government all law, whether private or public, operates in the public sphere, as it is defined, maintained and protected by the state, we described in great detail the way toxic tort attempts to govern the environment. This saw us dealing with many complexities, including the fact that the common law, while working within the broad norm of legal-political government, to do with maintaining and enhancing appreciable civil peace, occasionally supplements this with its own understanding of morality, to do with individual responsibility. Another such complexity concerned the way in which toxic tort has its own, intermittently eccentric, relations with science, whereby toxic tort courts will sometimes revere science for its solidity of evidence and sometimes dismiss it because it cannot do enough to reach the court's 'commonsense' threshold, itself a product of an array of factors.

In saying these things, we are judging the common-law government of the environment, particularly toxic tort, a modest success. It is helping the other aspects of the legal-political government of the environment and climate change (politics, the state, the economy, interests, science, and aesthetics) to meet the goals of (a) maintaining and enhancing appreciable civil peace and (b) protecting the environment. To say it is a modest success might seem to be damning the common law with faint praise, but it is not. As we said in Chapter 2, trying to control the ever shifting balance between the benefit of humans, towards as wide an appreciable spread of civil peace as possible, and the maintenance of nature is an imperfect endeavour if ever there were one. To be moderately successful is, in this light, as much as can be hoped for. Given that common-law calculations are 'pragmatic and situational' (O'Malley 2000: 477–78) and that common-law strategies are centuries old and have their own peculiar criteria of what the law should seek to achieve, particularly by way of making individuals accountable for their behaviour, it would be churlish, we feel, to expect it to do more than it does by way of protecting the environment. As we noted earlier in the chapter, toxic tort is a form of law which can help improve the environment generally only by improving the particular environment involved in each case. In its own way, then, this form of law is emblematic of a feature we keep pointing to – the legal-political government of the environment and climate change is not government by religion-and-morality, it is not government by private moral conscience, it is not government by lofty ideals, and its capacity to deliver on its norm of achieving an ever wider spread of appreciable civil peace is not limitless. In line with this, the legal-political government of the environment and climate change is never easy, never straightforward, and never complete or perfect.

5 Law's role in the legal-political government of the environment

Part 2: the public-law role

Introduction: the boundary between the common law and public law

Much of the furniture for this chapter has already been put in place. We have not only argued that law operates alongside politics in a form of government born in the sixteenth and seventeenth centuries – which also features sovereignty, state, interests, economy, aesthetics, science, and morality – we have also emphasized the way in which law, whether private or public, operates in the public sphere (defined as the sphere that both fosters legal-political government *and* is simultaneously fostered by it), where it plays a major role in the calculation and mitigation of various risks. In this chapter, consistent with this emphasis, we will deal, in more detail than we have to this point, with two prominent, interlinked entities, both of which are simultaneously contributors to and protected by the public sphere: public law and public interests.

Before we describe how the chapter is to be divided, we think it necessary to consolidate the distinction we have essayed in previous chapters between private law (especially the common law) and public law. We discussed the role of public law in Chapter 2, where we showed how it dovetails with politics in making legal-political government work. In Chapter 4 we made clear that private law (especially the common law) is built around individuals and their concerns where public law is built around the concerns of legal-political governments and the territories and populations they rule. And we stressed, also in Chapter 4, that both public and private law operate within the public sphere and, as components of legal-political government, contribute to it. In saying these things, we suggested that the boundary between these two forms of law is nowhere completely clear cut.

In the spirit of this last remark we would perhaps not be as formal about this boundary as are Sean Coyle and Karen Morrow, but nonetheless we are sympathetic to their attempt to describe the way it is usually drawn, particularly when it comes to governing the environment:

> Private law is characterized as the domain of individual entitlements, understood as a pattern of horizontal relationships between individuals.

Public law is depicted as a set of vertical relationships between individuals and the state. These latter exist, broadly speaking, as a set of limitations on the existence and scope of private rights, and controls upon their exercise. Such controls are deemed essential and desirable on a number of grounds, both in the context of distributive projects and projects aimed at increasing overall welfare.

(Coyle and Morrow 2004: 3)

To put the matter as simply as we can, each of private law and public law attempts to govern the environment in its own particular way – as this chapter will further explain – and while the differences between them are important, at the end of the day both forms are components of the same system of government and so both are pursuing the same goal (the maintenance of appreciable civil peace).

In the first main section we will describe environmental law as a specific form of public law. In the second section we will expand upon this by discussing the way environmental law operates in the public sphere, which will see us concentrating upon the way in which the notion of public interest has become important in environmental law cases, not just because of its role in helping to secure official legal recognition for advocates of environmental protection (a form of recognition known as 'standing'), but also because of its role in keeping the law firmly within the tradition of legal-political government. In the third section we will focus on some points of articulation between, on the one hand, environmental law and, on the other, politics, the state, and morality, which will see us discussing some existing legal analyses in a bid to further emphasize the importance of the idea of the public legal conscience to the effective legal government of the environment. And in the fourth section we will extend into the domain of public law our discussion of the importance of the notion of risk to the government of the environment and climate change.

Environmental law as public law in the public sphere

Coyle and Morrow (2004: 108) trace the birth of statutory environmental law back to the use of the common law in the nineteenth century to deal with the 'Industrial Revolution and its side effects'. Through the common-law jurisprudence of this period, they argue,

Property came to be seen as bounded in principle … The principles involved often related (roughly speaking) to conceptions of harm and wellbeing; yet the courts articulated such conceptions in terms of the notion of 'natural use': the extent of an owner's entitlement to use property in his own way was conceived as depending on the nature of the environment over which the rights were exercised.

(Coyle and Morrow 2004: 109)

In other words, while the common law identified principles to cover nature's well-being and harms to it (what would later be called, as we made clear in Chapter 3, the environment's well-being and harms to it[1]), it could not, because of the very way in which it operates, address these principles directly. Instead, it devised different strategies to deal with them indirectly.

One such strategy was the rule of 'natural use', which was first employed in the case of *Rylands v Fletcher* (1868). This 'natural use' rule was developed to deal with 'acute pollution problems'. It did so by applying strict liability for 'non-natural use', this being the term for 'dangerous activities' (Coyle and Morrow 2004: 121–22). Reasoning in terms of 'natural use' thereby 'allowed the courts a certain amount of latitude in deciding what activities would fall under the rule and gave them opportunity to temper the application of the rule in order to ensure that a satisfactory accommodation could be arrived at between conflicting individual and societal interests' (Coyle and Morrow 2004: 122).

Another indirect strategy was the common law of nuisance, which was developed to resolve disputes between 'property owners whose lands were adversely but indirectly affected by the activities of their neighbours' (Coyle and Morrow 2004: 111). Before the Industrial Revolution the law of nuisance was called on only to adjudicate a 'fairly narrow range of disputes', but the dramatic changes produced by the introduction of mass industrial production 'brought about a more complex and fragmented society, with conflict beginning to emerge between land uses which were increasingly at variance with each other' (Coyle and Morrow 2004: 112).

With its reliance on different strategies, the common law ended up applying a wide range of standards, of both well-being and harm, depending on the nature of the damage or interference involved and on the character of the locality where the nuisance was alleged (Coyle and Morrow 2004: 115). This was further complicated by different courts' use of the notion of 'reasonableness': 'Resort was made to the concept of reasonableness, both in term of the plaintiff's expectations and the defendant's activities as a rationale for legal interference' (Coyle and Morrow 2004: 116). For instance, in *Tipping v St Helens Smelting Company* (1865) the House of Lords ruled that the very tangible, physical contamination of Tipping's agricultural land by the acidic by-products of the defendant's copper smelting was unreasonable, whereas in *Sturges v Bridgeman* (1879) the defendant was ruled not to have caused unreasonable interference in the use and enjoyment of the plaintiff's property because the property was in an industrial area: 'Industrial areas, where nuisance was at its worst, obtained a lower degree of protection than rural environments where problems were fewer' (Coyle and Morrow 2004: 115).

Perhaps the best that can be said for the common law in these circumstances

---

1 Coyle and Morrow do not make the distinction between 'the environment' and 'what would later be called the environment'; we are including it in parentheses here because we think it an important distinction.

is that it highlighted a serious problem, one which only legislation could begin to address:

> Thus it is arguable that, at common law, pollution is regarded at least to some extent as a socially undesirable phenomenon. As such, it is a short step to recognise the need for regulation of polluting activities in the wider public interest, if only on the pragmatic grounds of forestalling the need for numerous expensive individual legal actions ... Where the common law ceases to be regarded as embodying a system of natural rights and duties based on rational principles, the development of systematic responses to generic social problems comes to be seen as the preserve of statute law.
>
> (Coyle and Morrow 2004: 130)

This is not to say, of course, that all common-law attempts to deal with pollution were in vain. In the course of the construction of the statutory road, particularly in the twentieth century, the work done by the common law continued to be felt. For example, the concept of

> statutory nuisance ... builds upon the common law, but is invoked in the public interest and is, in all other respects, a creature of public law proper. The concept of statutory nuisance was prominent from the inception of modern public health law: it appeared in a number of guises in the *Public Health Act* 1875, which remains in large part the template for the modern law in this area.
>
> (Coyle and Morrow 2004: 126)

How environmental law operates as a form of public law

Our overriding concern in this section is not to discuss the most important legislative attempts in different countries to protect the environment – this has already been done, and done well, by others (see especially Bell and McGillivray 2008; Fisher 2010; Godden and Peel 2010; Holder and Lee 2007; Kubasek and Silverman 2013; Lee 2005; Sands et al. 2012; Scott 2009). Rather, our main concern is to show that by operating in the public sphere, the most important contribution of environmental law to the legal-political government of the environment and climate change has been to formalize and highlight the public interest of the environment itself (on this matter we rely to some extent on Goodie and Wickham 2002). In this way, public environmental law has disrupted the law's tendency to give primacy to human individuals and their interests (a tendency particularly marked in private law, especially the common law).

Ewald argues that while individual interests may be the primary focus of judicial proceedings, the law's role in all modern governmental projects (what we would call the projects of legal-political government) is to deal with the interests of national populations (or sometimes those of international populations, as is the case with climate change). In Ewald's way of putting the matter (1986: 50–52), the state is organized to deal with ever-expanding, ever-shifting sets of 'collective interests', which occasionally must be aligned with particular individual interests. In this way, for him, collective interests are not simply the sum of individual interests.

While it has its origins in nineteenth-century forms of private law (which always focused on individual interests), as a body of public law which aims to protect the environment per se, environmental law is very much a product of the twentieth and twenty-first centuries, that is, of the period since the environment became an object of legal-political government. The environment's interests, therefore, are the interests of the state, which, as we keep saying, are the interests of the entire population of any law-and-politics country, interests captured by our definition of civil peace: the widest possible appreciable spread of peace, security, well-being, and prosperity among the humans within the territory being governed. So, when we use the formulation 'the public interest of the environment itself', we are not thinking primarily of a timeless, universal entity which exists well beyond the existence of human beings. While in a strictly scientific sense and in a strictly religious or metaphysical sense the environment might be thought of as a timeless, universal entity, these are not the main drivers of environmental law. This is because the law per se, in legal-political government, is not the law of the timeless universe, it is the law of civil peace, the law, that is, of the widest possible appreciable spread of peace, security, well-being, and prosperity among the humans within the territories being governed. As such, public environmental law is concerned with the long-term well-being of the environment precisely in being concerned with the long-term well-being of the humans within the territories ruled by legal-political governments. As we will argue in more detail in the following chapter, the law – indeed, the entire enterprise of legal-political government – is not, by its traditions, concerned with projects with a scope beyond this, though we will contend (in both the present chapter and the next) that law and politics are sometimes drawn into the idea that protecting the environment is some sort of timeless, universal moral quest, a quest which stretches beyond the boundaries of the territories ruled by legal-political governments. We remind the reader that while legal-political government always leans towards an anti-perfectionist understanding of the goals and aspirations of governing on a large scale, it does not always avoid perfectionist thinking about that which it attempts to govern.

In a very practical sense, for public law to be able to protect the public interest of the environment it had to develop a means to represent 'the environment' in court. It did this, eventually, through the official form of legal recognition we mentioned earlier, 'standing'. Principles of standing in environmental law – which had their origins in the common law but which were reformulated in the context of the growth of statutory environmental regimes – are concerned with determining whether a particular organization can act in court as an advocate of the public interest of the environment. These principles involve more than calculations about the representativeness of any organization's membership. They especially involve calculations about its expertise and its capacity to work with governments. In this way, within the confines of the system of legal-political government, questions of standing are at the heart of the law's work in protecting the environment; such questions are thus questions within the public sphere, they are questions about how public law works within this sphere to help legal-political government to protect the environment, in line with its goal of maintaining the widest possible appreciable spread of peace, security, well-being, and prosperity among the humans within the territories being governed.

However, before public environmental law could begin to operate in this way, the stranglehold of one particular common-law rule – a rule which was established by one particular case, *Boyce v Paddington Borough Council* (1903); hence it is known as the rule in *Boyce* – had to be broken. The rule in *Boyce* was used, until at least the late 1970s, in all Australian and English cases in which private persons sought injunctive or declaratory relief regarding activities undertaken for the public benefit. The rule effectively limited the environment's public interest, inasmuch as it stipulated that people and organizations acting to protect public spaces from harm would only be allowed standing in a particular case if they could show that in addition to there being a breach of a public right in that case, there was also (a) an interference with their private rights or (b) that the breach of the public right had caused the potential plaintiff to suffer 'special damage', over and above that which had been suffered by the public in general. In the Australian federal jurisdiction this meant that an individual, group, or organization would be granted standing only if they could convince the Attorney-General of the day (the nation's chief law officer) to grant them his[2] fiat. But in a catch-22 situation, Attorneys-General were reluctant to grant their fiat to private citizens in cases involving matters of public interest. This situation was criticized in a report issued by the Australian Law Reform Commission in 1985 (Australian Law Reform Commission 1985: 54–61). In particular, the Commission observed that, 'The record of the Attorney General in granting fiats is not a particularly generous one', a problem compounded by the fact that 'the rights of control which he possesses are extensive and denial of a fiat by him is not open to challenge in the courts' (Australian Law Reform Commission 1985: 61).

In this crucial period the Australian Conservation Foundation (ACF) and other non-government organizations worked hard to present themselves as obvious advocates of the public interest of the environment. But they kept bumping into the rule in *Boyce*. In 1980, for instance, when the ACF mounted a High Court challenge against the way in which an environmental impact process had been applied in assessing an application to build a tourist resort on the Great Barrier Reef, the Court found not only that the public duty

---

2 All Australian Attorneys-General in the period under consideration were men.

created by the Act was 'not owed to any particular person or persons', but also that the ACF did not have standing to enforce the public duty (*Australian Conservation Foundation Inc v Commonwealth* 1980: 524). The common law thus set the bar for the 'public' component of 'public interest' impossibly high, thereby creating an atmosphere against the idea itself. 'Public interest' was seen to be too nebulous, especially when compared to the concrete qualities of an individual litigant with identifiable legal rights. In the same case, one of the judges, Justice Gibbs, loosened the rule in *Boyce* such that, he suggested, it would have been enough for the ACF to have demonstrated a 'special interest' in the subject matter of the actions. However, the ACF was unable to demonstrate such a 'special interest' to the court:

> I would not deny that a person might have a special interest in the preservation of a particular environment. However, an interest, for present purposes, does not mean a mere intellectual or emotional concern ... A belief however strongly felt, that the law generally, or a particular law, should be observed, or that conduct of a particular kind should be prevented, does not suffice to give its possessor *locus standi*.
> (*Australian Conservation Foundation Inc v Commonwealth* 1980: 530)

The 1985 Australian Law Reform Commission, mentioned above, finally helped to change this mood, such that by 1987 an Australian Federal Court judge, Justice Wilcox, was able to acknowledge the importance of organizations like the ACF and thereby acknowledge, albeit indirectly, that entities like the environment could have their own public interest advocates in court:

> Much of the progress of mankind has been achieved by people who have sacrificed their own material interests in order to champion ideals against fierce resistance. The recent Australian experience is that, in cases where ideologues have been able to gain access to the court, cases have been hard fought and professionally conducted. I illustrate this point by referring to ten reported cases, involving diverse issues arising in different parts of Australia and a variety of plaintiff groups.
> (*Ogle v Strickland* 1987: 58)

In 1988, just one year on, the words of Justice Wilcox and two other Federal Court judges – Justices Davies and Gummow – show just how fast attitudes had changed:

> The term 'interest' has long been an expression used in the law with respect to parties so as to require an involvement with a case greater than the concern of a person who is a mere intermeddler or busy body ... The necessary interest need not be legal, proprietary, financial or other tangible interest. Neither need it be peculiar to a particular person.
> (*Re: United States Tobacco Company and Australian Federation of Consumer Organisations Inc. and: The Minister for Consumer Affairs; The Trade Practices Commission and Australian Federation of Consumer Organisations* 1988: 86)

This was further confirmed the year after, 1989, in a major case before the Federal Court, *Australian Conservation Foundation v Minister for Resources*, when Justice Davies ruled that

> [P]ublic perception of the need for the protection and conservation of the natural environment and for the need of bodies such as the ACF to act in the public interest has noticeably increased, and is demonstrated by the growth of the ACF itself since the time of the [1980] *ACF case*.
>
> (1989: 73)

Effectively, then, in the course of the 1980s the notion of public interest was 'redefined', something which was made clear in 1993 when the Public Interest Advocacy Centre held a summit under the title 'Redefining the Public Interest'. The organizers discovered, as one of the participants, Simon Rice, observed, that the very manner in which the meeting was conducted demonstrated that the 'redefinition' was already complete, not long into the 1990s:

> The 'redefinition' which was the aim of the conference was never explicitly addressed; no session of the day was set aside for that issue. Rather, the day's agenda itself reflected the range of interests and perspectives that could contribute to a definition of the public interest. During the day an idea developed of the many aspects of 'the public interest', and of the competition among interests any of which could claim to be part of the public interest.
>
> (Rice 1993: 192)

While we will shortly discuss a 1990s setback for the role of the public interest, this comment from Rice is nonetheless a good way to portray the public interest of the environment: a range of interests loosely gathered around the single object – 'the environment' – to function as a single 'public interest'.[3] More specifically, courts now recognize the legitimacy of non-government organizations operating within the public sphere in matters deemed to constitute the public interest of the environment. In line with this, organizations-as-advocates of the public interest of the environment, alongside organizations-as-advocates of public health, now regularly take advantage of the new judicial openness. In this new atmosphere, non-government organizations have had more success in their legal work on behalf of the environment than individual citizens ever did under the old arrangements. This has been due largely to their expertise and to their ability to organize and to lobby governments and private firms; that is, by the fact that they are keen

---

3 At this juncture we remind the reader of a point we made in Chapter 2, that the notion of public interest does much of its legal work as rhetoric. In environmental law, this is to say, it gives environmental lawyers rhetorical opportunities to persuade a court to act on an established legal position *and* it allows the court to be so persuaded.

to use both law and politics when dealing with legal-political government. These organizations have come to be regarded, by the general public as well as by courts, as formal guardians of the environment, mostly working with governments and private firms, but sometimes against them.

In a 1994 case before the New South Wales Land and Environment Court – to determine whether a decision by the Richmond River Council to allow residential subdivision of a block of land would have a detrimental impact on native fauna, especially koalas – the judgment of Justice Stein made clear that public interest in the environment owed its new strength to public law:

> The notion of public interest litigation has been gaining ground in Australia over the last decade. The concept derives from principles of public law rather than private. The development springs from the increasing access of individual members of the public and groups to approach the courts and seek to enforce aspects of public law. This is particularly so in the area of environmental law where many New South Wales statutes include open standing provisions enabling 'any person' to seek to enforce breaches of the law.
>
> (*Oshlack v Richmond River Council* 1994: 2)

The judge also praised the efforts of the individual environmental advocate (Oshlack) for persevering in his attempts to have environmental statutory provisions understood and enforced:

> The basis of the challenge was arguable, raising serious and significant issues resulting in the important interpretation of new provisions relating to the protection of endangered fauna. The application concerned a publicly notorious site amidst continuing controversy. Mr Oshlack had nothing to gain from the litigation other than the worthy motive of seeking to uphold environmental law and the preservation of endangered fauna. Important issues relevant to the ambit and future administration of the subject development consent were determined ... These issues have implications for the Council, the developer and the public.
>
> (*Oshlack v Richmond River Council* 1994: 7)

However, it was not all clear sailing for the public-law idea of the public interest of the environment. It still had to be defended well into the 1990s, inasmuch as the 1998 High Court appeal against Justice Stein's judgment revived the spirited legal debate about whether the public interest of the environment can be represented by individuals and organizations in court, a debate which focused on the way the presiding judge had calculated this matter. On the one hand, Justice Kirby argued that Justice Stein had been correct in the way he had recognized the public interest (*Oshlack v Richmond River Council* 1998: paras 136–37), but Justice McHugh, on the other hand, argued that Justice Stein had in fact departed from accepted legal practice.

For Justice McHugh, a more precise definition of public interest was still required (*Oshlack v Richmond River Council* 1998: para 72). In making his argument, Justice McHugh cited Professor Paul Gewirtz as an authority:

> Judicial power involves coercion over people, and that coercion must be justified and have a legitimate basis. The central justification for that coercion is that it is compelled, or at least constrained, by pre-existing legal texts and legal rules, and by legal reasoning set forth in written opinion. From this perspective, the exercise of judicial power is not legitimate if it is based on a judge's personal preferences rather than law that precedes the case, on subjective rather than objective analysis, on emotion rather than reasoned reflection.
>
> (Gewirtz 1996: 1025)

Justice McHugh, this is to say, was concerned that the original judgment relied on 'nothing more than the social preferences of the judge' (*Oshlack v Richmond River Council* 1998: para 75). Chief Justice Brennan was of a similar mind, but the two of them were ultimately in the minority. By a majority the Court accepted that the public interest in the original case was calculated in a legally acceptable manner.

We think this to be a crucial judgment. To put it in the terms of our argument, in this case the High Court accepted (albeit not unanimously) that the public law operated as it should always operate: on the basis of its public legal conscience, not on the basis of a private moral conscience. We will say more about this matter in the next section.

Environmental law, politics, the state, and morality

We said in the introduction to this chapter that this section would see us discussing some existing legal analyses in a bid to emphasize the importance of the idea of a public legal conscience to the effective legal government of the environment. In heading into this discussion we wish to reiterate a point we made a short while ago, albeit only as an aside: that within the framework of legal-political government successful recourse to environmental law usually involves a direct combination of law and politics. We are not sure that this is what the first of two texts we will focus on here – Coyle and Morrow (2004) – is saying when its authors argue that environmental law is concerned 'with the intrinsic worth of aspects of the natural environment, and not merely an instrumental concern with current social conditions' (Coyle and Morrow 2004: 2) and nor are we sure it is what the second text – Fisher (2003) – is saying when its author, Douglas Fisher, argues that, 'Environmental law is one of those areas of the law that is identified by its underlying philosophy and by its subject matter rather than by the nature and source of rights and obligations that sustain it' (Fisher 2003: 1).

However, because we are so keen to stress that environmental law, as a branch of law operating within legal-political government, must be fundamentally concerned with the maintenance of appreciable civil peace, we intend reading these two texts as if this is indeed what they are saying. As such, we will take Coyle and Morrow's point about environmental law including a commitment to the 'intrinsic worth of aspects of the natural environment' as something which has to be *reinforced by* the 'instrumental concerns with current social conditions'. In this way, while we regret their use of the word 'merely' in describing such instrumental concerns, we take Coyle and Morrow to be allowing, even encouraging, a political awareness of current conditions as part and parcel of environmental law's commitment to the intrinsic worth of aspects of the natural environment (for we think it is the case that all *effective* law-and-politics environmental law protections are formulated in the context of current political conditions).

Similarly, we do not think Fisher's proposition that environmental law is 'identified by its underlying philosophy and by its subject matter' is inconsistent with our position. While it is certainly the case that for the legal-political way of thinking 'underlying philosophies' are not always conducive to the widest possible appreciable spread of peace, security, well-being, and prosperity among the humans within the territory being governed, we realize that this does not mean that they are necessarily unhelpful. Rather, it means that, by the traditions of legal-political government, decisions about whether they are helpful or unhelpful will not be left solely to their advocates. For this way of governing, it is always the case that: (a) 'the state cannot possibly undertake its core functions of pacification and security unless it can decide for itself, "without internal impediment", what can be publicly expressed, or just who can own what' (du Gay 2012: 403, quoting Dunn); and (b) the state must have this supreme power *on behalf of all other actors*, '"the state carries, and must carry, the authority of its own subjects' will and choice to make that judgement on their behalf and to act, decisively, upon it"' (du Gay 2012: 403, quoting Dunn). We will show below that Fisher is very much aware that the protection of the environment can only be successfully pursued in law-and-politics countries when each of law, politics, and the state is allowed to play its role.

This is the case, too, with other discussions offered by these analysts. For example, when Fisher extends one of his observations – that 'the legal framework within which the management of the environment takes place comprises a complicated set of interlocking rights, duties, powers and liabilities of diverse kinds' (Fisher 2003: 1) – into a two-pronged argument that (a) 'Traditional concepts of law as a regime for setting and enforcing standards of behavior and decision-making have been found wanting' and (b) 'The mechanisms for a legal system – powers, liabilities, rights, and duties – have failed to confront the complexity of environment' (Fisher 2003: 24), we take him to be saying that the law should not attempt to act without the support of politics and the state in seeking to protect the public interest of the environment. And when he speaks of the need to 'infiltrate the system' with 'ideas of responsibility for achieving outcomes – sustainable development, for instance' (Fisher 2003: 24), we take him to be saying, as we will discuss in more detail shortly, that 'sustainable development' is not a perfectionist goal but a carefully crafted strategy by which a combination of law, politics, the state, business interests, and pragmatic environmental activists can take steps to protect the environment without giving ground to the idea that the earth must be returned to its perfect natural state.

In other words, we think that both texts are suitably aware that the legal-political government of the environment and climate change is never easy, never straightforward, and never complete or perfect. We think that the three authors involved know that environmental law, which has itself been operating for less than fifty years, is part of a system of government that has proved itself capable, over the course of nearly four hundred years, of maintaining appreciable civil peace. In offering these remarks in support of their efforts, we share their frustration that environmental law is too often having to catch up to polluters and is not always as clear in its direction as it might be. As such, we treat them as allies in thinking it best to judge environmental law against a very broad picture of what the law has on its plate, in working as a part of legal-political government, and we trust they would agree with our somewhat crude point that statutory environmental regimes in law-and-politics countries like Australia (we might here call them civil-peace countries) are much more effective than those in countries in which appreciable civil peace is not the order of the day.

In saying all this we are obviously further defending the idea that if legal government of the environment and climate change is to remain within the legal-political tradition, a public legal conscience must take precedence over private moral consciences. To take this issue further, let's start with Fisher's excellent description of 'the environment' as an object in need of protection:

> Even if the environment was originally taken to be an area within locationally and spatially identifiable boundaries, it has now assumed a meaning sufficiently wide to include not only such areas but also anything and everything within those areas. Moreover for many it now implies a value rather than a mere description: for example the need to use, conserve or protect the environment.
>
> (Fisher 2003: 24)

Here we take Fisher to be treating the task of 'conserving or protecting the environment' as a pragmatic one. By this way of thinking, law and politics support the state as it ensures that the 'value' of the environment remains a matter for debate and negotiation, that it is not hijacked by the private moral campaign to return law and politics to the perfectionist path from which they supposedly strayed in becoming the mainstays of legal-political government.

This brings us back to the notion of sustainable development, which Fisher says (2003: 56) is about both 'the development of resources and the protection of the environment'. Coyle and Morrow (2004: 203–05) are equally confident that sustainable development – as a means of justifying and directing the course of present development *and* guaranteeing the sustainability of the natural environment – can 'enhance the carrying capacity of the resource base'. In this way, we suggest, if environmental law is promoting sustainable development in the manner that both Fisher's book and Coyle and Morrow's book say it is, then it is definitely pursuing what we call the normative trajectory of legal-political government – seeking to maintain the widest possible appreciable spread of peace, security, well-being, and prosperity among the humans within the territories being governed. Another scholar, Maarten Hajer, confirms this in describing sustainable development as the product of a 1980s environmental discourse which attempted to reconcile 'economic restructuring with environmental care'. Hajer also argues that the 'alternative and conceptual language' of the sustainable development strategy effectively marginalized the then-dominant radical environmental agendas by promoting pragmatic, technical solutions to environmental problems in place of 'confrontational' posturing (Hajer 1995: 95). He goes still further (1995: 100) in saying that sustainable development has demonstrated its capacity to 'get environmental issues out of the periphery of politics (as conservation issues) and sought to link them to core – i.e. economic – concerns'.

Almost as soon as it was formulated as a strategy, in the late 1980s, sustainable development was embraced by many governments around the world (Hajer 1995: 9, 101). In Australia, Fisher explains (2003: 353), the federal government led the way with its 1992 *National Strategy for Ecologically Sustainable Development* (Commonwealth of Australia 1992b). This was followed, in the same year, by the *Inter-governmental Agreement on the Environment* (Commonwealth of Australia 1992a) which committed federal, state, territory and local governments to the principles of sustainable development as the key means of governing the environment (we will return to this *Agreement* in the following chapter, when we discuss some doubts raised in some early 1990s debates about whether the notion of sustainable development was strong enough to do the work it was being asked to do). These principles included aspirations (like biodiversity and intergenerational equity), market incentives, direct legal liabilities (such as polluter pays rules), and procedural directions aimed at smoother decision making in the planning process. Fisher is clear (2003: 351–70) that, as one would expect when so many different interests come to the same table, the arrangements were something of a hodgepodge. Nonetheless, he argues, even though some of the desired outcomes were pursued through law while others were pursued through politics and still others through business, the strategy worked. Of course it did not work smoothly or completely – giving us another opportunity to stress that the legal-political government of the environment and climate change is never easy, never straightforward, and never complete or perfect – but it did manage to give the coalition involved – the state, public law, politics, business interests, and environmental forces – a shared (if shaky) sense of unity.

At very least the strategy served to provide a relatively positive answer to an important question Fisher had asked earlier in his book:

> While an ethic of conservation [or sustainability] may be justified in various ways, the essential conundrum for a legal system, either international or national, is to find a form in which it can be meaningfully expressed, recognized or enforced. Is there such a grund norm? ... Is there, in other words, a right of environment and can the legal system recognize and protect it?
>
> (Fisher 2003: 48)

We can say, then, that in using sustainable development to tie together different modes of governing the environment, the Australian approach described by Coyle and Morrow and by Fisher has proved to be more valuable than has the perfectionist approach that wants nature to rule over humans in a perfectionist manner, as it was supposedly intended to do: 'It is one thing to postulate the ethical foundations of nature conservation and environmental protection. It is quite another to translate these ideologies into a form and substance that are both meaningful and enforceable within a legal system' (Fisher 2003: 75).

In its practicality, Fisher contends (2003: 101), the sustainable development strategy has encouraged a situation whereby environmental policy 'has become an integral part of the fabric of environmental law in Australia as well as in other jurisdictions ... Policy in this sense is a formal declaration of environmental values. When it is an element of legislation – as it increasingly is – it is more than a policy direction ... [it is an] environmental charter.'

The continued role of risk

The fact that public law has been more effective in governing the environment than private law, inasmuch as it can deal with large-scale governance where private law cannot, does not mean, of course, that the notion of risk is less important to public law than it is to private law. Indeed, the effectiveness of the public-law government of the environment and climate change has spawned even more ways to think about and address risk. In the twenty-first century the intersection of public environmental law, politics, the state, private law, insurance, and the science of the environment (ecological science) has produced a new instrumental approach to the government of the environment and climate change, an approach in which risk figures heavily (Fisher 2003: 353–54, 363; Harremoes et al. 2002: xv; Peel 2005: 18; Steele 2004).

By this instrumentalist approach, the health of the environment and the health of humans are mapped and evaluated through specialist calculations of the 'risk-weighted consequences' of particular activities, especially those to do with the use of natural resources (Peel 2005: 22). Paul Rutherford (2000: 56) argues that 'regulatory ecological science does not so much describe the environment as both actively constitute it as an object of knowledge and, through various modes of positive intervention, manage and police it'. In this way, the government of the environment and climate change turns on techniques for calculating risk as diverse as epidemiology (McMichael 2001: 22), environmental impact assessments (Peel 2005), and protocols for 'scoping' corporate greenhouse gas emissions (McGrath 2007: 220). It is these types of calculation of risk that allow those officially charged with managing the environment (including those tribunals and courts which are called on when disputes about the environment reach a certain level) to make objective and scientifically informed calculations about its well-being.

This is not, however, how every analyst sees these changes to the management of the environment. Aynsley Kellow, for example, argues:

> Environmental issues inevitably entail questions of both morality and practicality. The very notion of 'pollution' is culturally dependent; it is a moral category which predates the environmental era, and environmental pollution derives from wider notions of pollution. Similarly risk and hazard are subjective qualities. There is no objective scientific basis for public policy in general and environmental policy in particular.
>
> (Kellow 1999: 12)

It should come as no surprise to the reader that we would put this point differently. Perhaps one could say that 'pollution', 'risk', and 'hazard' were subjective entities based in moral discourses *before* the rise of the legal-political government of the environment, but a key feature of this type of government, as we discussed in detail in Chapter 2, is that it has chased morality off the public stage and into the private realm, where people are free to pursue their moral and religious convictions as they see fit, with the important proviso that they are not permitted to do so in a manner that threatens civil peace. So, under legal-political government 'pollution', 'risk', and 'hazard' are definitely not subjective categories. They have been made into public governmental categories, which are as objective as such categories can be – they are the calculations of scientific experts, statisticians, clinicians, economists, and, since the late 1990s, non-expert participant citizens (Steele 2001).

In this, our position is much closer to Rutherford's proposition (1999: 116) that risk has become an idiom through which the environment is made 'thinkable' and 'amenable to political deliberations'. For him, 'the relationship of society to the natural environment is conceived in terms of the language of security and risk; ecological hazards and insecurity must be addressed by putting in place behaviours that minimize risk'.

We might usefully develop this insight by dipping again into the governmentality literature (as we did in Chapter 2), this time in the form of an argument from two of Foucault's most significant Anglophone followers, Peter Miller and Nikolas Rose, about how entities are made objects of government:

> [K]nowing an object in such a way that it can be governed is more than a purely speculative activity: it requires the intervention of procedures of notation, ways of collecting and presenting statistics, the transportation

of these to centres where calculations and judgements can be made and so forth. It is through such procedures of inscription that the diverse domains of 'governmentality' are made up, that 'objects' such as the economy, the enterprise, the social field and the family [and the environment] are rendered in a particular conceptual form and made amenable to intervention and regulation.

(Miller and Rose 1990: 5)

Beck makes a related point in emphasizing that,

[R]isks open the opportunity to document statistical consequences that at first were always personalized and shifted onto individuals. In this way risk de-individualizes. Risks are revealed as systematic events, which are accordingly in need of general political regulation ... A field for corresponding political action is opened up: accidents on the job for instance, are not blamed on those whose health they have already ruined ... but are stripped of their individual origin and related instead to the plant organization, the lack of precautions, and so on.

(Beck 1999: 51)

Before we head to the chapter's conclusion, we should acknowledge that in placing points from Beck cheek-by-jowl with points from governmentality thinkers like Miller and Rose, we are deliberately side-stepping the supposed differences between the Beck approach and the Foucauldian governmentality approach. Where Beck's broad 'risk society' thesis is pointedly epochal – his determination to divide the history of human societies into grand eras and to focus on the most recent of these eras, which he calls 'late modernity', is clearly the source of his 'discovery' of 'risk society' (Beck 1992, 1999) – those working within a governmentality framework (for example, Dean 1999; Ewald 1991, 2002; O'Malley 1999a, 1999b, 2004; Weir 1996) tend to be more interested in the operation of particular techniques and technologies for governing risk than in whether or not these techniques and technologies add up to a new form of society or to a new morality. In this sense, we lean more towards the governmentality position, but inasmuch as we think there are at least hints of epochalism in the very idea of an era of 'governmentality' (see Wickham 2006, 2008), we prefer not to be drawn into this dispute. Instead, we wish to continue to borrow points from both camps, for we regard both as stores of useful insights into the operations of different forms of government. Indeed, we will borrow another such insight in the conclusion.

Conclusion

This chapter has added an account of the role of public law to the book's overall picture of the legal-political government of the environment and climate change. We think this addition looks quite a lot like this:

[A] delicate affiliation of a loose assemblage of agents and agencies [loosely operating as] a functioning network. This involves alliances formed not only because one agent is dependent upon another for funds, legitimacy or some other resource which can be used for persuasion or compulsion, but also because one actor comes to convince another that their problems or goals are intrinsically linked, that their interests are consonant ... This is not so much a process of appealing to mutual interest as of what Callon and Latour term *interessment* – the construction of allied interests through persuasion, intrigue, calculation or rhetoric.

(Miller and Rose 1990: 9–10)

Miller and Rose are not here describing the government of the environment and climate change. Nonetheless, what they have to say captures very succinctly the way in which politics and public law operate in seeking to protect the environment. This is especially true of the way in which, often deploying the notion of sustainable development, politics and public law attempt to look after the public interest of the environment precisely as they look after their own interests, which of course has them also looking after those of the state, and those of other key actors, like private and public economic entities (for example, firms and government finance units).

Our answer to our book's key question – how do law and politics, entwined in various complex relations with sovereignty, the state, morality, religion, economy, aesthetics, and science, attempt to govern the environment and climate change? – is, all this makes plain, now broader than it has been at any point so far. In offering four more chapters since posing the question in Chapter 1, we have now answered it in terms of the basic operation of the various components of legal-political government, in terms of the unusual mix of aesthetics and science that brought the environment to life as a particular object of legal-political government, and in terms of the two forms of law most directly charged with protecting the environment – private law (especially common law) and public (statutory) law. In doing these things we have taken pains to show that these elements perform their 'governing the environment' work in concert with the work done by the other elements of this type of government, particularly the economy, and to show that all of them always keep a weather eye on their core task of maintaining the widest possible appreciable spread of peace, security, well-being, and prosperity among the humans within the territory being governed.

In the course of the next chapter our already complex answer to the key question will need to be made even more complex, for there we will be examining the way legal-political government is trying to deal with the threat of climate change, which some have argued is the greatest challenge the environment has ever served up to the world's governments.

6   The legal-political government of the threat of climate change

Introduction: adjusting the climate of the climate change debate

In 2007, in one of his campaign speeches, the soon-to-be Prime Minister of Australia Kevin Rudd offered a newsworthy slogan about the threat of climate change: 'The greatest moral challenge of our generation' (Rudd 2007). While these words undoubtedly had and still have wide appeal, we nonetheless think they are extremely misleading. The way we see it, climate change is not primarily a moral challenge, or an economic challenge, or a social challenge. For us, the threat of climate change is primarily a governmental challenge, a challenge which the legal-political system of government is very much attempting to meet.

In the spirit of our approach, we prefer to turn away from the slogans of politicians and towards Machiavelli's advice, first published, in *The Prince*, in 1532,[1] about what a wise ruler should do to prepare for the vicissitudes of fortune, or nature as it is often called today:

> I judge it to be true that Fortune is the arbiter of one half of our actions, but that she still leaves the control of the other half, or almost that, to us. And I compare her to one of those ruinous rivers that, when they become enraged, flood the plains, tear down the trees and buildings, taking up earth from one spot and placing it upon another; everyone flees from them, everyone yields to their onslaught, unable to oppose them in any way. And although they are of such a nature, it does not follow that when the weather is calm we cannot take precautions with embankments and dikes, so that when they rise up again either the waters will be channeled off or their impetus will not be either so disastrous or so damaging. The same thing happens where Fortune is concerned: she shows her force where there is no organized strength to resist her; and she directs her impact there

---

1 *The Prince* was widely circulated in manuscript form for a number of years before it was formally published.

where she knows that dikes and embankments are not constructed to hold her.

(Machiavelli 1979: 159)[2]

In other words, while climate change is undoubtedly a big challenge for the legal-and-political system of government, it is nonetheless a challenge *within* the scope of this system's one norm: maintaining the widest possible appreciable spread of peace, security, well-being, and prosperity among the humans within the territories being governed. So, in the face of the threat of climate change, legal-political governments around the world (as well as those parts of the UN and other international bodies which operate along legal-political lines) will act in the manner of Machiavelli's wise ruler. That is, they will act so as to ensure, to the best of their abilities, that 'the impetus' of these calamities 'will not be either so disastrous or so damaging' as they would be were governments not as experienced and as well equipped as are legal-political governments. Just as the British government does in dealing with the threat of houses on unstable cliffs sliding into the sea (without letting morality trample over the interests of law, politics, sovereignty, the state, the economy, aesthetics, and science), and as the Australian government does in dealing with the threat of bushfires, and as the US government does in dealing with the threat of tornadoes and hurricanes, and as the Japanese government does in dealing with the threat of earthquakes and tsunamis, and as do all other legal-political governments in dealing with whatever slings and arrows climate change fires at their territories. In this way, legal-political government will use all the governmental resources and tactics it has developed over several hundred years *and* it will deploy its capacity for 'trial and error' in a bid to find replacements for those governmental resources and tactics which prove inadequate.

Of course, in including the Japanese government's response to earthquakes and tsunamis and the US government's response to hurricanes we are reminded of our earlier point that these two governments did not, by the best practice standards of legal-political government, do very well in helping those affected by the 2011 earthquake and tsunami in northern Japan and by Hurricane Katrina in New Orleans. It is important that we say again that legal-political government fails on many occasions, a reality softened, though only slightly, by the fact that it usually admits to its failures and does what it can to fix them and/or not repeat them (which is not always enough).

While some extreme activists may wish to turn the challenge into an eschatological event – for them, climate change, because they regard its primary cause to be human involvement, is an opportunity to judge humans and to

---

2 By including this quote from Machiavelli, we are not suggesting that climate change was a concern of governments in his era, though it is worth noting that some authors do discuss the effects of climate change on European governments in the sixteenth and seventeenth centuries (see, for example, Parker 2013).

find them wanting – this is not, in and of itself, a concern for legal-political governments. Such governments will not seek to prevent people from thinking that the anthropogenic component of climate change issues makes climate change into 'the greatest moral challenge of our generation', though of course if any of people holding more extreme views than this should threaten civil peace by any actions born of their moral convictions, legal-political governments will become more than concerned about them. Just as such governments have always done – it is in their DNA – when private moral or religious convictions boil over in any particular law-and-politics country to the point that they threaten large-scale civil violence, the government of that country will act decisively to quell the threat. As we said in Chapter 2, borrowing a point from Stephen Holmes, the threat of civil war is the *'summum malum'* for this type of government, 'the uttermost evil to be avoided at all costs'. Climate change is a large and growing concern for legal-political governments around the world, but it is nowhere near the *'summum malum'* for any of them, though of course this does not, nor should not, lessen their concern about, for example, the devastating impact of rising sea levels and/or tidal inundation in Kiribati and Bangladesh, where thousands of people have been displaced.[3]

With our position on the government of the threat of climate change firmly stated, we can now set out the way in which the chapter is to be divided. In the first main section we will further discuss the way the threat of climate change is being understood as a problem by legal-political governments. In the second section we will discuss some of the legal, political, economic, and scientific strategies being employed by legal-political governments to tackle the threat of climate change, at both national and international levels. In the third section we will introduce and thoroughly discuss the 'precautionary principle' as the fulcrum of a separate strategy, one which combines legal, political, economic, and scientific elements such that governments can make the notions of sustainable development and risk into a single package, which they hope will be used more widely in dealing with the threat of climate change. In both of these sections we will deal with the various strategies in a positive light, focusing (mostly) on what the strategies are trying to do and/or what they are actually doing. In the fourth section we will turn in the other direction and present a range of negative assessments, from within the legal-political system, of all of the strategies dealt with in the third section. In this way, the balance of the chapter will reflect the 'trial and error', 'never smooth sailing' nature of legal-political government.

---

3 Saul et al. (2012: 191–92) argue that climate change is 'likely to aggravate the scarcity of basic resources' in such countries, which might in turn aggravate 'political and economic instability', though, importantly for us, they add that this is unlikely to lead to civil war or even to significant conflicts.

How is the threat of climate change being understood as a problem by legal-political government?

Alexander Zahar, Jacqueline Peel, and Lee Godden summarize the governmental challenge of climate change in these terms:

> Climate change is a global problem touching all nations, yet one that manifests itself in innumerable local forms that have their own immediacy in the domestic context ... [In confronting it] 'top-down' international rule making is complemented by 'bottom-up' national and regional contributions to the shaping of legal concepts in the field ... [It] may be thought of as more than just another environmental problem, but instead as one that envelopes and affects all others. It demonstrates ... the impact that humans have on the natural world, and it challenges the systems humans have or are able to devise to limit their impact on the environment as no environmental problem has done before.
>
> (Zahar et al. 2013: 1–2)

We agree with them only in part. Yes, climate change is a big challenge for legal-political governments around the world, and, yes, it has provoked a growing mixture of local, national, and international legal regulations. But where they think it is a problem like no other problem before it, we think legal-political governments are in fact treating it as a new, very large problem which shares many features with problems they have dealt with before or are still dealing with. In this way, just as it did with projects like the rebuilding of Europe after the Second World War and is still doing with projects like the rebuilding of the financial system in the wake of the Global Financial Crisis (where it is having to deal with its own failures more regularly than it and the people being governed would like), in being a pointedly anti-perfectionist system, legal-political government is working with whatever tools are to hand, trying not to get ahead of itself, mixing pragmatism with principle, and, in the manner of its own science component, is not attempting to find ultimate truths but is using more and less sophisticated 'trial and error' processes to find the best-for-now solutions, which are then themselves being instantly subjected to the same process.[4]

The example of the ongoing rebuilding of the global financial system is especially telling in this regard. In trying to identify the best workable ways of managing the threat of climate change, legal-political governments are drawing on everything they know from previous attempts to address similar problems and adapting them as best they can to what they are now confronting. They face myriad difficulties, both in trying

---

4 As Jan McDonald (2010: 2) describes the role of science in the government of climate change, 'climate science has not been able to provide the level of certainty that policy and decision makers seek in order to inform their decision making, nor is it likely to do so'.

to figure out what is going wrong and in trying to fix at least some of it. The difficulties include (to name just a handful): unreliability of data from previous projects; lack of co-operation between the governments of different law-and-politics countries; the weaknesses of international agreements and international law (mentioned in Chapter 1 and to be discussed in more detail later in this chapter); and obfuscation and feet dragging on the part of some only-because-we-have-to participants at both the public governmental level and the private commercial level (for more on this last phenomenon, see Hulme 2009).

In short, the threat of climate change is being understood by legal-political governments as a major problem to be dealt with in the way that legal-political governments deal with major problems. A huge amount of managerial, intellectual, legal, political, and financial resources are being dedicated to the threat, as one would expect, but of course this will not change the point we keep making: legal-political government is never easy, never straightforward, and never complete or perfect, as our fourth section (the 'negative assessment' section) will demonstrate. Nonetheless, for those who value the widest possible appreciable spread of peace, security, well-being, and prosperity among the humans within the territories being governed, this is the best that can be done, as our second and third sections (the 'strategies as they are meant to work and as they sometimes are working' section) will demonstrate.

Some legal, political, economic, and scientific strategies being employed by legal-political governments to tackle the threat of climate change

We have decided, for ease of presentation (and consumption), to divide our examples of governmental strategies into only two groups, with each group to be discussed in a separate sub-section. The first group will deal with predominantly legal strategies. The second group will deal with three types of predominantly political strategies: (a) those that are primarily political; (b) those that combine politics with economics; and (c) those that combine politics with science. Of course, we are not claiming anything like comprehensiveness for our coverage of the different strategies. All we are trying to do in this section is to give a flavour of the different methods and devices legal-political governments are employing in trying to make sense of the threat they face from climate change and in trying to combat that threat (for more detailed and more wide-ranging treatments of these matters, see Bonyhady and Christoff 2007; Bonyhady et al. 2010; Brunnée et al. 2012; Burns and Osofsky 2009; Cole 2011; Giddens 2009; Helm and Hepburn 2009; Hulme 2009; Lord et al. 2012; Prins et al. 2010; Saul et al. 2012; Stallworthy 2009; Vanhala and Hilson 2013; Zahar et al. 2013).

Legal strategies

As Elizabeth Fisher notes (2007: 14–16), the role of the law in governing risks such as climate change is not clear cut, in part because the law sometimes tries to align itself with scientific and moral rationalities. In the words of Zahar et al. (2013: 2), 'From a legal point of view, climate law does not fit a pre-existing category. It does not slot easily into any developed branch of environmental law.' And in writing about the interface of law and science, Sheila Jasanoff (1995: xiv) is of a similar mind: 'Advances in the realm that is conventionally labeled "technological" require the readjustment of existing behaviours, institutions and relationships'. To tackle the complexities of climate change, this is to say, legal principles are being rethought and reworked. This is true, for instance, of principles concerning the limit of individual property rights, of principles concerning the legal responsibility of governments for the well-being of the natural world, and of principles for determining the type of conduct that constitutes harm.

In particular, in dealing with climate change cases, courts are still trying to work out the appropriate balance between national and international interests. In Australia, for example, 'Courts … have on occasion … [interpreted] the domestic principle of "ecologically sustainable development" as requiring attention to the global aspects of climate change, for example by balancing "geographically narrower concerns" with the "broader public good of increasing the supply of renewable energy"' (Zahar et al. 2013: 4).

In this context, deference to the significant economic impact of any measure mandating action to mitigate and abate greenhouse gas emissions (GHG emissions), for example, along with the scientific difficulty of calculating the effect of such gases on climate change, has constrained the will to introduce direct legal regulation. As was the case in the 1980s and 1990s with the environment in general, this has left litigation by environmental advocates and non-government organizations to carry much of the burden of asserting and articulating the legal obligation to take action on climate change (Peel 2007: 96; Peel and Osofsky 2013: 20–22; significant cases in Australia and the US – most of them in the twenty-first century – include: *Aldous v Greater Taree City Council* 2009; *American Electric Power Co. v. Connecticut* 2011; *Australian Conservation Foundation v Latrobe City Council* 2004; *Charles & Howard Pty Ltd v Redland Shire Council* 2008; *Coalition for Responsible Regulation v EPA* 2012; *Drake Brockman v Minister for Planning* 2007; *Gippsland Coastal Board* 2008; *Gray v Minister for Planning* 2006; *Greenpeace Australia Ltd v Redbank Power Company* 1994; *Kivalina v Exxonmobil Corp.* 2009; *Massachusetts v EPA* 2007; *Minister for Planning v Walker* 2008; *Myers v South Gippsland Shire Council (No.2)* 2009; *Northcape Properties Pty Ltd* 2008; *Queensland Conservation Council v Xstrata Coal Pty Ltd* 2007; *Taralga Landscape Guardians* 2007; *Wildlife Preservation Society of Queensland* 2006). Peel and other scholars, like Bonyhady and Douglas Fisher, think this strategy is eventually likely to produce favourable results, for at least three reasons: first, Australia has a

body of specialist environmental courts that are experienced in assessing scientific calculations of technological risk; second, these courts operate within a
paradigm that is cognizant of the principles and methodology of ecologically
sustainable development; and, third, the courts share the widespread aspiration to achieve intergenerational equity, that is, the bid to ensure that present
problems are not fobbed off onto future generations (Peel 2007: 96; Bonyhady
2007: 19; Fisher 2003: 351–54).

The Australian courts' reputation for 'independence from government and
capacity to develop legal principles of broad application' has also encouraged non-government organizations to initiate climate change litigation (Peel
2007: 96). The strengths and limitations of this Australian litigation have
been the subject of much debate amongst academics and lawyers (see, for
example: Bonyhady and Christoff 2007; Lin 2012; Lyster 2007; Peel 2007;
2008; Peel and Osofsky 2013). Tim Bonyhady (2007: 19) goes so far as
to argue that Australian environmental jurisprudence has started to turn
the principles of ecologically sustainable development into substantive law,
though he also acknowledges that the key cases divide equally on the question
of whether climate change is occurring. In any event, it might be that climate
change litigation since 2006 is serving as a warning to governments that they
should take more decisive action on climate change. Peel argues that despite
the uncertainty, expense and the 'ad hoc nature of court proceedings', the
litigation in question,

> may have substantial, flow-on benefits for environmental law in
> general, particularly by increasing the rigour of EIA [Environmental
> Impact Assessment] practices and clarifying the obligations of envi-
> ronmental decision-makers with respect to achieving the goals of ESD
> [Environmentally Sustainable Development] … Ultimately, the most
> significant contribution of climate change litigation in Australia may
> be putting the GHG issue firmly on the government and business
> policy agenda in a way that provides momentum for stronger regulatory
> action.
>
> (Peel 2007: 103)

Of course, as with all legal thinking, thinking on climate change law is
subject to contest and resistance. As such, the optimism of Peel, Bonyhady,
and Fisher is not shared by everyone, as we will demonstrate in our 'negative
assessments' section. Nevertheless, the outcome of one particular case fully
supports their position. This is the 2006 case of *Gray v Minister for Planning*,
which became known as the Anvil Hill case. The judge, Justice Pain, found
that the environmental impact assessment of a new open-cut coal mine at
Anvil Hill, which was to supply coal to a power station, must consider the
greenhouse gas produced by the burning of the coal as well as the emissions
produced by the mine itself. She also ruled that the greenhouse gas emissions
produced by burning the coal would 'contribute to climate change/global

warming, which is impacting now and likely to continue to do so on the Australian and consequently NSW [New South Wales] environment' (*Gray v Minister for Planning* 2006: 100). In this case Justice Pain is insisting that important decision makers 'take proper account of the principles of inter-generational equity and the precaution' (*Gray v Minister for Planning* 2006: 114–15).[5]

Political (and economic and scientific) strategies

In this sub-section, as promised, we will describe some instances of primarily political strategies, some in which politics combines with economics, and some in which politics combines with science (though we feel the need to add yet again that each of these three overlaps with the other two and each of them overlaps with the legal strategies).

The main examples of primarily political strategies we wish to high-light are the United Nations Framework Convention on Climate Change (UNFCCC) of 1992 and the Kyoto Protocol of 1997. These strategies sought to establish 'a framework of objectives and institutions with ongoing respon-sibility for bringing climate change under control' and 'reporting and com-pliance systems that have helped to harmonise the measurement of GHG emissions in developed states (but not yet in developing ones)' (Zahar et al. 2013: 3).

The logic of these two closely related strategies is to seek common ground and to devise rules which should, in theory, apply to all countries committed to the Convention and the Protocol: 'As all countries emit GHGs to some extent, the responsibility for anthropogenic GHG emissions is common, but the UNFCCC clearly placed real responsibility, both in the sense of historical blame for global warming and in the sense of now doing most about it, on the developed countries' (Campbell et al. 2010: 168). In line with this, the Convention and the Protocol have mostly been used to try to make interna-tional rules effective within 'domestic systems'. Zahar et al. (2013: 3) agree that the international rules under discussion 'have overwhelmingly pertained to reporting on country-level climate policies and measures, accounting for national GHG emissions … and compliance checks on national institutions charged with carrying out these functions' and in this way have been 'more strictly applied to developed than to developing countries'. Nonetheless, they say of these rules that, in working alongside 'the fundamental principles set down in the UNFCCC (the precautionary principles, common but differenti-ated state responsibility, the interests of future generations, etc.), they may

5 In a US case of even greater significance, *Massachusetts v EPA* (2007), the Supreme Court, in creating 'direct pathways' for federal climate change regulation across the country, helped to legitimize the issue of climate change 'in the eyes of the general public' and forced companies to give greater consideration to their greenhouse gas emissions and even to the possible envi-ronmental consequences of their normal business practices, especially their planning (Peel and Osofsky 2013: 157).

be said to constitute a rudimentary international (procedural) climate law' (Zahar et al. 2013: 3).

Our first example of a strategy which combines politics and economics is the Clean Development Mechanism (CDM), which operates as part of the Kyoto Protocol. CDM projects are funded by developed countries in exchange for 'emission reduction certificates', which they receive only if they have successfully off-set an agreed-upon amount of the excess emissions they produce. In the developing world CDM projects have two complementary objectives: to help developing countries to produce less carbon-intensive products and services; and to offset carbon emissions in developed countries. If a CDM works as it is meant to work, it should, argue Zahar et al. (2013: 201), produce quantifiable GHG reductions. In considering this matter, Campbell et al. (2010: 174) propose that, at least in some ways (but certainly not in others, as we will show later), the mechanism is working as it is meant to work: 'The UNFCCC has met with great success in generating support for carbon trading from very powerful public and private interests' and the number of CDM projects registered 'has grown beyond all initial expectation'.

Zahar et al. offer a good account of the way in which CDM projects try to extend the reach of the legal-political government of climate change into 'weakly governed' states. In this way, a CDM project attempts to 'carve out a small, internationally supervised space ... in which to introduce reforms that are beyond the [local] government's capacity'. The success of such attempts to govern climate change in this way, they add, depends on whether it is practically possible 'to defend the borders of this space throughout the project's lifetime in order to maintain the integrity of the emission reduction credits' (Zahar et al. 2013: 212).

Our second, closely related political-economic example is the development of various carbon trading schemes, or emissions trading schemes as they are often called. One form of these schemes, known as 'cap-and-trade' schemes (to be defined shortly) have been favoured by developed nations over direct regulation of GHG emissions, largely because 'it is difficult to put into place standards that can be readily adjusted to give effect to progressively lower emission reduction targets' (Zahar et al. 2013: 170). All emissions trading schemes allow the setting of mandatory targets, but provide flexible mechanisms for meeting those targets, thereby giving industry time to make the technological transition to less carbon-intensive operation (Zahar et al. 2013: 176).

Cap-and-trade schemes operate by setting an upper limit – the cap – on overall GHG emissions for each participant in each scheme. Each participant is allocated rights to emit GHGs up to an agreed-upon limit. As for the trade component of cap-and-trade, if the participant exceeds its agreed-upon level of emissions, it either submits to sanctions (sanctions which themselves are also formally agreed upon when the scheme is established) or it can buy enough permits to cover its actual emissions from any other participant in the scheme which has kept its emissions below its allocation, thus leaving it with permits to trade. Emissions can also be offset by technologies that facilitate carbon capture. Such schemes are supposed to encourage reduction in overall GHGs over time by progressively limiting their caps on total emissions. In this way, at least in theory, provided that the 'emission units' are equivalent, the schemes can operate trans-nationally as well as nationally (Zahar et al. 2013: 173–74).

Because the schemes are designed to aid 'the mitigation of global warming', they are seen as 'an important part of a fundamental "civiliza- tion" of markets' (Campbell et al. 2010: 165). They have been described by David Campbell, Matthias Klaes, and Christopher Bignell (2010: 164) as an extension of welfare economics: '"carbon trading" under the 1997 Kyoto Protocol (KP), is the first major exercise in what we will call the "new global welfare economics", in which the welfare economics that has dominated national economic and social policy in the twentieth century has been extended to a global scope and scale'. Emissions trading schemes are now so widespread that they are definitely the main game in town: 'regardless of what comes out of international climate negotiations, emissions trading is likely to remain a dominant regulatory approach, albeit perhaps not in the top-down fashion envisaged by the Kyoto Protocol' (Zahar et al. 2013: 187).

Our one example of a strategy which combines politics and science con- cerns the necessary development of models and protocols for the calculation of the aggregate impact of GHG emissions on the world's climate:

> Through its exhaustive, continuously revised compilations of science, which involve the international engagement of thousands of scientists and extensive global promotional and lobbying activities, a UN agency, the Intergovernmental Panel on Climate Change (IPCC), has managed to establish a 'scientific consensus' that 'dangerous anthropogenic inter- ference' (DAI) with the global climate is taking place … The response to this is to seek to limit anthropogenic GHG emissions, and therefore atmospheric GHG concentration, at a level which will curb the warming effect.
>
> (Campbell et al. 2010: 167)

As this suggests, the main achievement of this marriage of politics and science is the production of a very specific target. This target, based ostensibly on 'IPCC calculations about the danger posed by various concentrations of GHG', is that there must be a 'reduction of global emissions such that warming will be limited to a 2°C increase over pre-industrial temperatures' (Campbell et al. 2010: 168). The 2°C target was first 'adopted by the European Union in 1996 (and reaffirmed in 2005)' and 'attracted general international support' at the Copenhagen summit in 2009 (Saul et al. 2012: 62).

Of course, reaching a consensus so as to produce a target, while undoubt- edly an important achievement in international politics, is not necessarily the same thing as reaching the target set by the consensus, a matter to which we will return in our 'negative assessments' section.

The precautionary principle as an element of the legal-political government of the threat of climate change

In Chapter 5 we discussed some of the ways in which the notion of sustainable development, which combines law, politics, economics, and science, has proved a useful tool for legal-political governments in dealing in general with environmental risks. In dealing with the specific threat of climate change, however, while the 'pre-existing general legal principles affirmed in the UNFCCC and the Kyoto Protocol include that of "sustainable development"' (Zahar et al. 2013: 4), the use of the notion of sustainable development came to be questioned, largely because it was seen to be not strong enough to do the work being asked of it in such challenging circumstances:

> The cumulative, climate-related consequences of our human activities threaten an overwhelming externality problem going to the root of the sustainability of human society itself and exacerbated by the greatly diffused impacts (and information deficits) that afflict production and consumption decisions.
>
> (Stallworthy 2009: 417)

As Zahar et al. put the matter:

> climate law requires a more powerful principle to respond to the larger environmental threat … [S]ustainable development *simpliciter* has a limited scope that does not necessarily extend beyond that of national interest. Globally sustainable development, by contrast, is the kind of legal principle that would fit the logic of the climate change problem. A public authority or government bound by the principle would be obliged to weigh global sustainability in decisions that have an environmental impact, even if no impact is expected within its own jurisdiction.
>
> (Zahar et al. 2013: 4)

In arguing this position, Zahar et al. are deploying the precautionary logic which informed Justice Pain's decision in the Anvil Hill case.

An important precursor to the concern about the notion of sustainable development occurred in Germany in the 1970s. In that country, the long-standing idea of precaution had been developed into a principle known by the compound German word *Vorsorgeprinzip*. 'Vorsorge is premised on ideas of taking anticipatory measures to avoid environmental damage, ensuring detection of dangers to human health and the environment, and acting before conclusive scientific evidence of harm becomes available' (Godden and Peel 2010: 240). These ideas became more widespread in the 1980s. For example, at the North Sea Conference of 1987, the notion of the 'precautionary principle' was considered as a possible basis for protecting the North Sea marine environment (Godden and Peel 2010: 240).

The shift from local to global was hurried along by the formal definition of the precautionary principle delivered at the 1992 United Nations Conference on Environment and Development in Rio de Janeiro, sometimes known as the Rio Summit:

> 'In order to protect the environment, the precautionary approach shall be widely applied by States according to their capabilities. Where there are threats of serious or irreversible damage, lack of full scientific certainty shall not be used as a reason for postponing cost-effective measures to prevent environmental degradation.'
>
> > (Godden and Peel 2010: 240, quoting the
> > *Rio Declaration on Environment and Development*)

The Rio event remains pivotal to the rise of the precautionary principle, largely because of this 'grand vision' definition. In other words, the definition, especially inasmuch as the Summit included it 'in two major multilateral treaties concluded at the Earth Summit – the *Convention on Biological Diversity* and the *United Nations Framework Convention on Climate Change*' – helped to give the precautionary principle the very high status it has now in the fight against climate change. It was because of Rio that 'it has become one of the highest profile principles of international environmental law, and an established element of domestic environmental law' (Godden and Peel 2010: 240).

We now turn away from the more formal aspects of the precautionary principle and towards its regular operation. It is important to note that the principle works with regard to only a certain class of risks. As Ewald argues, it is not intended that the principle should apply to 'all risk situations' or potential harms, only to those for which there is both 'scientific uncertainty' and 'the possibility of serious and irreversible damage' (Ewald 2002: 283–84). Nevertheless, the widespread acceptance and institutionalization of the principle marks what a number of commentators have described as a paradigm shift in the political and legal understanding of the scope of environmental government (de Sadeleer 2002: 91, 229; Ewald 2002: 276; Hajer 1995: 27, 67; Peel 2005: 21). As Nicholas de Sadeleer puts it (2002: 91), 'The question is no longer merely how to prevent assessable, calculable, and certain risks, but rather how to anticipate risks suggested by possibility, contingency, plausibility, probability'. In a similar vein, Hajer (1995: 67–68) argues that to understand environmental government in precautionary and 'anticipatory' terms is the 'antithesis' of 'remedial environmental politics'. As a basis for framing legal obligations, he contends, the precautionary principle was conceived as an antidote to the 'react and cure' form of regulation, which was increasingly criticized as inadequate in the face of those environmental harms with uncertain and potentially global effects (Hajer 1995: 26).

The principle thus operates with what Ewald calls the 'logic of decision to act or not to act' and in doing so it 'extends the field of subjective responsibilities because it focuses on the act of decision'. He argues that even though it only 'makes sense before the decision is taken', it inevitably also provides a basis for sanctioning particular decisions and assigning legal responsibility for them, largely because it insists that a decision can be 'judged not only by what one should know but also by what one should have or might have suspected' (Ewald 2002: 287).

Douglas Fisher contends that the precautionary principle continues to have significant consequences for the ways in which environmental law is interpreted and for the scope of its application. Its application requires and allows environmental decision makers to think beyond their own immediate jurisdiction, to think in terms of an environment which is less confined by a material space and more multi-dimensional in its operation. He argues that 'the incorporation of sustainability places much greater emphasis upon long-term implications. This has come to mean the long-term implications not only for the land and its immediate environment but also for the wider environment' (D. Fisher 2007: 216).

Also writing on the way the principle has widened the horizons of environmental government, Ewald (2002: 282, 288) argues that since the 1980s we have been confronted by 'the return of disasters', most of which are attributed to human agency. These disasters are, he says, defined by their devastating and often irreversible consequences, by uncertainty as to when and how the disaster will manifest, and uncertainty as to how to counter and contain the harm it produces. Uncertainty is now the order of the day.

As uncertainty has taken on this increased significance, social scientists have taken an increased interest in how we construct and act on the uncertainty itself (Ewald 2002; Furedi 2009; Hanekamp et al. 2005; Jasanoff 2010; Shaw 2010). Their research suggests two principal and interlinked ways in which uncertainty disrupts established means of understanding and managing the environment. First, uncertainty confounds the authority of science (Ewald 2002: 273–74). Second, as both Frank Furedi (2009: 206) and Jasanoff (2010: 238) contend, it undermines confidence in the future. By Ewald's way of framing this matter, since the nineteenth century the scale and probability of risks has been determined 'against a background of scientific expertise', but this very expertise is proving inadequate in the face of the uncertainty associated with climate change. In this way, uncertainty has intensified 'in the light of an even newer science [ecological science] ... one that is consulted less for the knowledge it offers than the doubt it insinuates' (Ewald 2002: 274).

While not disagreeing with Ewald on this point, Jasanoff (2010: 238) takes a slightly different tack in arguing that the dynamic that produces doubt is not simply a product of what science predicts but also a product of the fact that science's account of the future does not align with 'established ways of understanding the human place in nature'. She proposes that because it is only through science that climate change has come to prominence, rather than through other more immediate media of human experience, its existence, its scale, its time-frame, and its predicted impact all remain hotly contested (Jasanoff 2010: 235).

In line with this, the uncertainties of climate change are also inextricably linked to its more insidious attributes, such as the latency of its manifestation, its apparent invisibility, its pervasiveness, and its likely irreversibility. In this context, Ewald concludes (2002: 282–83), it is not surprising that the spectre of climate change causes us to be fearful, or that climate change itself is characterized as a crisis or disaster (we will return to this theme in the following sections).

It is no less surprising, then, that climate change is such a challenge to legal-political government. Uncertainty is no friend of this type of government. To take just one aspect of the problems created by the new level of uncertainty, climate change and global warming disrupt established jurisdictional bases for determining legal obligations and asserting legal rights, which have come to be conceived until recently in terms of the 'immediate past and near-term future' (Jasanoff 2010: 242). Whereas the risks associated with conventional accidents and with certain lifestyles and diseases can be identified, contained and offset, the uncertainty of climate change denies this level of control and makes it impossible 'to articulate a standard of conduct that everyone must observe' (Ewald 2002: 297).

In Australia, as elsewhere, the adoption of the precautionary principle followed quickly on the heels of the Rio Summit. As we discussed previously, later in 1992 the federal government and all the state governments jointly formulated and agreed upon the *Inter-governmental Agreement on the Environment* as one of the tenets of environmental policy (E. Fisher 2007: 130). The *Agreement*'s definition of the principle largely follows the Rio definition, but is less grand in its language:

> Where there are threats of serious or irreversible environmental damage, lack of full scientific certainty should not be used as a reason for postponing measures to prevent environmental degradation.
>
> (Commonwealth of Australia 1992a)

The *Agreement*, Douglas Fisher notes (2007: 217), was obviously meant to commit all levels of government in the country to formally adopt the principle so as to allow it to extend the scope and reach of legislation. But in fact the principle has not been uniformly implemented either across or within jurisdictions. Indeed, Elizabeth Fisher (2007: 131–32) argues that not only has it not been fully implemented in Australia, it has come to be understood as a 'discretionary and flexible' guideline, not as a strict rule. She goes on to contend that even in courts and tribunals conducting judicial and merit review of decisions taken on the environment, which is where the principle's meaning and scope in Australia has largely been determined, the principle has made only a modest impact, producing little more than an acknowledgement of the 'common sense' of precaution in decision making (E. Fisher 2007: 132).

Godden and Peel (2010: 263) are of a different mind. They think that an acknowledgement of the need for prudence in decision making is very significant in itself, that this 'theme of cautious decision making' has encouraged 'a wide range of precautionary measures', including waiting on further scientific evidence before proceeding with a development, imposing an ongoing environmental auditing obligation, and accepting measures to mitigate potential environmental harm (Godden and Peel 2010: 263). This is consistent with Douglas Fisher's point (2007: 217) that the judicial and merit review cases potentially extend the scope and reach of legislation to incorporate the longer term and geographically more extensive impacts of greenhouse gas emissions.

Clearly, the success or otherwise of the precautionary principle remains a matter for debate, and this is true not just for Australia but for all law-and-politics countries. As is the way with legal-political government, with so many competing interests encouraged to participate in the politics of any issue, there is no sure route forward on matters such as the value of this particular strategy. In the spirit of this mode of understanding the legal-political government of the environment and climate change, we move now to our 'negative assessments' section. As we said earlier, legal-political government is always a 'trial and error', 'never smooth sailing' affair.

Negative assessments of strategies for governing the threat of climate change

If the above assessment of the operation of the precautionary principle in Australia is nearer the 'trial' end of our 'trial and error' formulation, this section deals with assessments nearer to the 'error' end. In other words, if the previous two sections dealt mainly in descriptions of some of the ways in which the different strategies described were realized, or partly realized, this section will deal mainly in descriptions of some of the ways in which those hopes were dashed, or partly dashed.

Within our sample of criticisms of strategies for dealing with the threat of climate change, the strongest criticisms are those aimed at the primarily political, political-economic, and political-scientific strategies, though some of the criticisms we will present on the precautionary principle strategy are nearly as strong. Because of the way the law operates, critical commentary in that field is usually quite restrained. We gave more than a few hints of this in a previous section, when we discussed the way the legal strategies are operating, but this characteristic is even clearer in the small cluster of criticisms we have chosen to represent negative assessments of legal strategies. We will present this cluster before we move to the criticisms of the various political strategies and the criticisms of the precautionary principle strategy.

The law cluster relates to a legal decision we outlined earlier – Justice Pain's decision in the Anvil Hill case to the effect that the GHG emissions produced by the burning of coal would 'contribute to climate change/global warming, which is impacting now and likely to continue to do so on the Australian and consequently NSW environment' (*Gray v Minister for Planning* 2006: 100). Some of the reporting and editorials on the decision, as highlighted by Farrier (2007: 205), were quick to claim that Justice Pain had ventured from the bench into policy making and/or that her decision was driven by an ideological commitment to shutting down the coal industry. More than this, the government most directly affected by the decision, the state government of New South Wales, attempted to neutralize the decision's political significance by portraying Justice Pain as an environmental radical and then by avoiding the consequences of her ruling (Farrier 2007: 205–07).[6]

We are keen that the reader does not get the wrong impression here. It is certainly not the case that this sort of criticism applies to all, or even a majority, of legal strategies for dealing with the threat of climate change. But there is at least some evidence that anthropogenic climate change has become something of an elephant in some courtrooms, where it is discussed only to doubt its existence (for a similar argument, see Lin 2012: 44). For example, Justice Dowsett, in the case of *Wildlife Preservation Society of Queensland* (2006: para. 72), summed up his understanding of the applicant's position in these terms:

> The applicant's concern is the possibility that at some future time, protected matters in Australia will be adversely and significantly affected by climate change of unidentified magnitude, such climate change having been caused by levels of greenhouse gases (derived from all sources) in the atmosphere ... The applicant's case is really based upon the assertion that greenhouse gas emission is bad, and that the Australian government should do whatever it can to stop it, including, one assumes, banning new coal mines in Australia.

The strongest criticisms of the primarily political strategies go to the heart of the system of legal-political government – the notion of sovereignty. In this way, the Kyoto Protocol and some of its mechanisms are criticized for trying to skate around the constraints of national sovereignty, especially when it comes to dealing with the strongest countries, like the US and China. Here is an example relating to the US:

> The United States' rejection of the Kyoto Protocol, the unambiguous mitigation targets of the Protocols commitment periods, and the economic recession that struck developed countries in 2009 (early in the

---

6 Nonetheless, the Anvil Hill case did prompt the government to introduce 'new State Environmental Planning Policy for mining activities requiring the inclusion of indirect emissions in further environmental assessments' (Peel and Osofsky 2013: 170).

Protocol's first commitment period) have not merely failed to substantially strengthen the climate regime over the years; they have moderated its demands.

(Zahar et al. 2013: 4)

As for China, Campbell et al., referring to it by the acronym PRC (People's Republic of China) and referring to the Kyoto Protocol by the acronym KP, argue that

the economic success of the PRC poses the main challenge to global climate change mitigation. The PRC recently replaced the USA as the world's largest emitter of anthropogenic GHGs ... the PRC currently is responsible for more than half the growth in emissions ... The global nature of the externality on which the KP strategy turns will then work against Annex 1 reductions [primarily those of advanced Western countries], making them pointless unless the PRC makes concomitant reductions.

(Campbell et al. 2010: 178)

Furthermore, they argue, it is unlikely that from a 'practical political' standpoint the Communist Party in China will abandon its ambitions for economic growth: 'political stability in the PRC' requires economic growth to raise the 'low living standards of the bulk of its population'. These political considerations 'outweigh the welfare gains from anthropogenic GHG reduction' (Campbell et al. 2010: 178).

Their thinking on this point is consistent with the remarks we made about China in Chapter 2, where we noted that it is a country with a genuine commitment to peace, security, well-being, and prosperity, but without a commitment to individual freedoms for its population. This means that its calculations about what constitutes peace, security, well-being, and prosperity for the Chinese people will be driven not by the people's involvement in the decision-making process but by the internal machinations of its ruling party, the Communist Party, which has made clear for some time that it believes China's fortunes as a powerful country rest heavily on rapid economic development.

This last point highlights the principal difficulty faced by international agreements about steps to mitigate the threat of climate change (and by all other such international agreements): even the most powerful international bodies, including the UN, do not have permanent sovereignty over any territory; they are reliant for their effectiveness on the sovereignty of their member states, especially the more powerful states. The fact that these powerful states have gained their power, at least in part, against the will of other powerful states means that it is highly unlikely that all powerful states will ever end up taking the same action towards achieving the goals of any international agreement, including those on climate change. This does not mean that they will never line up on the same side of a statement of aspiration, like stopping world hunger, delivering world peace, or reducing the threat of climate change. Rather, it means that the politics of international supremacy can accommodate agreements on aspirations far more easily than they can produce meaningful coordinated actions. As Martti Koskenniemi (2001) demonstrates in his book on the rise and fall of international law between 1870 and 1960 (and as we hinted in Chapter 1), it is one thing to draw on organic communitarian thinking to build a consensus around the idea of an 'international community', but it is quite another to have that community impose its supposedly shared interests on all its member states *over and above* their own sovereign interests.

The criticisms of political-scientific strategies tend to be quite technical, often dealing with things not done rather than with things done badly, as is the case with the failure of the procedures which were meant to verify the Clean Development Mechanism. Zahar et al. argue (2013: 216) that verifying the carbon benefits of CDM has proved onerous and difficult, mainly because a CDM project can never be fully transparent. One upshot of this absence of transparency, they point out, was a long-running 'perversion of the system'. This 'perversion', which lasted until 2011, had to do with the fact that firms which manufactured the non-greenhouse refrigerant gas HFC-22 actually had to use the powerful greenhouse gas HFC-23 in manufacturing it. The firms earned Certified Emission Reductions for their efforts, which meant that they were 'making more money from producing more greenhouse gas than they would have otherwise' (Zahar et al. 2013: 221; for more detail on this 'perversion' of the system, see: Campbell, Klaes, and Bignell 2010: 180–81). This particular problem has now been resolved, but 'Questions remain about the extent to which CDM reductions are real, measurable and verifiable' (Zahar et al. 2013: 227).

From this example the reader can probably guess that the strategies which have come in for the most criticism are the political-economic strategies, in particular the emissions trading schemes, precisely because, as Gwyn Prins and Steve Rayner argue (2007: 274), such schemes will not work in the top-down manner imposed by the Kyoto Protocol; instead, they 'must be built – like all markets – from the bottom up'. In line with this, some criticisms aim at the way the schemes were conceptualized in the Kyoto Protocol: 'The extent to which it was initially envisaged that the common responsibility would eventually "evolve" to impose substantial burdens on non-Annex 1 countries [developing countries] is a matter of contentious diplomatic history' (Campbell et al. 2010: 172). Campbell et al. go on to make clear that they are using the phrase 'a matter of contentious diplomatic history' as a euphemism:

Though the UNFCCC and the KP make no sense unless it was believed non-Annex 1 countries would eventually commit to reductions, and that the KP would make such commitment easier because those countries had been drawn in to carbon trading which would greatly reduce the cost of those reductions, there is nothing in either the Treaty or the Protocol to lead one to think this will happen; quite the contrary is in fact the case, for the entire history of UNFCCC diplomacy has been one of extending a permission to developing countries to emit as much as they choose.

(Campbell et al. 2010: 172)

This is of course the direct opposite of what the UN continues to believe is possible for its carbon reduction schemes, even in the face of evidence that the Kyoto Protocol was too ambitious, to say the least. For example, in 2007 Christiana Figueres, who has been Executive Secretary of the UNFCCC since 2010, observed:

'The KP is limited in its global emission reduction target, in its time frame, and in the countries that participate. Given the scale of the climate challenge, the KP can only be seen as the preamble of an extended and enhanced effort, which should continue to rely heavily on market mechanisms in order to manage the costs.'

(Campbell et al. 2010: 189, quoting Christiana Figueres)

Campbell et al., obviously not sharing Ms Figueres's relentless optimism,[7] go so far as to argue that carbon trading is doomed to failure because the 'technologically global nature of the problem' means that the targets of the developed (or Annex 1) countries and those of the developing (or non-Annex 1) countries are 'mutually dependent'. Because of the 'continuing absence of targets for non-Annex 1 countries' and the fact that it is not possible 'to have confidence in the implementation of the CDM (and therefore the global carbon market), there is no rational way in which any individual targets can be set' (Campbell et al. 2010: 187). Zahar et al's assessment is only slightly less bleak; they say that there is little in the 'market's core seller-buyer relationship that provides an incentive to undertake environmentally sustainable projects' (Zahar et al. 2013: 224).

Turning now to some criticisms of the precautionary principle strategy, Furedi (2009: 205–06), after noting that environmentalists were among the first to 'devalue probabilistic thinking' in favour of possibilistic thinking, argues that the precautionary approach is deeply possibilistic, relying much more on 'intuition' than on calculation, and in this way fosters unnecessary 'worst case' alarmism. While he acknowledges (2009: 206) that threats such as global terrorism and climate change are 'never fully knowable', he thinks probabilistic risk assessment is still the best basis for policy responses to these threats, which leaves him concerned by the increase in 'possibilistic-driven worst case policies' in law-and-politics countries: 'A possibilistic

---

7 Nonetheless, her optimism is widely shared by many people across the world, with efforts continuing unabated in many countries to develop national and regional carbon markets (Cole 2011).

interpretation of problems works to normalize the expectation of worse possible outcomes and fosters a one sided and fatalistic consciousness of the future' (Furedi 2009: 207).

Furedi's doubts about the precautionary approach to policy making are shared by Chris Shaw (2010), whose analysis of policy making on global warming identifies a governmental tendency to resist decisive, positive action based on probabilistic approaches to more complex and indeterminate aspects of environmental uncertainty. Instead, he argues, precaution is used to justify a go-slow approach to environmental government which does nothing but leave the danger of climate change to an alarming future.

Shaw contends (2010: 104; 115–16) that while climate science is often paraded as the basis of the narrative that demands that the planet's warming be limited to a 2°C increase over pre-industrial temperatures, the climate scientists themselves do not accept the premise of this narrative: that there is scientific consensus on the limit of dangerousness. He argues (2010: 106) that when this narrative is coupled with the precautionary approach its political appeal is heightened; by making the future look more dangerous, 'go slow policy prescriptions and the ecological modernization of existing technologies' look more feasible than they should. Mark Stallworthy (2009: 414) suggests that this criticism is often levelled against what many see as an 'addiction to target setting' in a bid to reduce GHG emissions.

Furedi (2009: 207–11) takes a similar line when he argues that the precautionary approach, in encouraging 'policies designed to deal with threats which are increasingly based on feelings and intuition rather than evidence and facts', actually 'inflates the power' of the indeterminate threat, whether it be terrorism or climate change, thereby justifying every 'precautionary act' and causing all do-something policies to be questioned or even abandoned (we will return to this theme in the following chapter).

Where Furedi's and Shaw's criticisms focus mainly on the governmental policy consequences of precautionary thinking, Hajer's negative assessment concerns the operation of science. Hajer argues that the precautionary principle has influenced the way in which scientists report their findings and possibly even influenced the conclusions they are prepared to draw from their data. Writing more than a decade before Furedi, he is adamant that in shifting the ambit of 'accepted statistical norms' the principle has 'altered the scientific practice and contaminated scientific discourse' (Hajer 1995: 67–68).

Conclusion

In this chapter we have built an account of the way in which legal-political governments have come to understand the threat of climate change – as a governmental problem and not a moral problem, as a problem they can deal with by developing strategies from what they have learned in dealing with other large-scale threats. We have described a range of the strategies being developed – some using law, some using politics, some using economics, some using science, and some using a new legal-political mechanism called the precautionary principle – and we have discussed a range of criticisms of their failings. In this way, we have added some much-needed micro detail to the climate change part of our answer to the book's key question: how do law and politics, entwined in various complex relations with sovereignty, the state, morality, religion, economy, aesthetics, and science, attempt to govern the environment and climate change?

In doing these things we have, in particular, provided some concrete evidence of the way in which the legal-political government of the environment and climate change involves a distinction we drew in Chapter 2. On the one hand, the 'appreciable' character of legal-political governments' commitment to civil peace, because it leads to constant debate by different interests, produces a very complex politics of the environment and climate change, one in which 'trial and error' is standard practice and in which it is impossible to ever be sure of the fate of any one position compared to the fate of all the others. On the other hand, in countries where governments lack this type of commitment, such as China, the politics of the environment and climate change are nowhere near so complex, where widespread debate is not encouraged, where a much more predictable set of interests is likely to be active, and the biggest difficulty facing anyone trying to assess the fate of any one position compared to the fate of all the others is to determine which position is being favoured in the ruling party, for its success is all but assured.

It has to be added that we have also provided some concrete evidence for another of our oft-repeated points: that legal-political government has a long record of serious failures, albeit one accompanied by a long record of trying to fix its failures.

7 Conclusion

Two case studies and some final thoughts

Summarizing the book

In introducing our book, we referred to it as a book about how the environment has emerged in the modern world as an object of the legal-political form of government. As the book progressed, we added many layers to this bare-bones description. We showed that the environment is many things to many people and that it is, in being many things, a source of life and pleasure, a source of death, destruction, and fear, and often both of these at once. We offered a comprehensive account of the way legal-political government operates (an account which gave roles to each of law, politics, sovereignty, the state, the economy, interests, morality, religion, aesthetics, and science) and we described the way in which a fortuitous combination of aesthetics and science first made the environment into an object of this system of government. From there, we took up the task of describing the particular role, within this system, of both private and public law in attempting to govern the environment. Finally, we put all of these components together and discussed the complexities of the response of the legal-political form of government to the threat of climate change.

We have made many points about the operation of legal-political government in general and the legal-political government of the environment and climate change in particular. We think all of them are important, but here we think it necessary only to restate what we regard as the seven most important points.

One, the legal-political system of government both relies upon and contributes to a strain of anti-perfectionist thinking about humans by which it rejects the necessity of trying to perfect humans, thereby rejecting the opposing idea of government by competing perfectionist visions, each determined to impose its own vision of good government on the others, whether that means forcing the followers of the other positions to submit to its particular form of perfection, or killing them.

Two, the legal-political system of government works with only one overriding norm: the pursuit of the widest possible appreciable spread of peace, security, well-being, and prosperity among the humans within each territory being governed (which we sometimes refer to as 'the maintenance of appreciable civil peace').

Three, the legal-political system of government is distinguished from other systems by its *appreciable* commitment to peace, security, well-being, and prosperity (or civil peace) among the humans in each territory being governed, which means that in countries ruled by a legal-political government, the law must be strong enough to temper politics in such a way that they work together in order that the great majority of the population of that country will have the opportunity to appreciate and exercise the individual freedoms that the maintenance of peace, security, well-being, and prosperity will have delivered to them, even if they choose not to take advantage of that opportunity.

Four, the legal-political system of government has replaced the stranglehold of the idea that all government must be driven by a conscience sourced in an external, supposedly timeless and universal morality, a conscience, that is, external to law and politics; in this way, legal-political government deals with the argument that government without a conscience is doomed to amorality and/or immorality by stressing the alternative normative compass set out above and by pointing to its own conscience of civil peace, a conscience which compels law and politics to drive government to pursue the widest possible appreciable spread of peace, security, well-being, and prosperity among the humans within each territory being governed.

Five, the legal-political system of government is not government by religion-and-morality, it is not government by private moral conscience, and it is not government by lofty ideals.

Six, the legal-political system of government is never easy, never complete, and never perfect (a point we have regularly reinforced by pointing out that legal-political government has a long record of serious failures, accompanied by a capacity and a determination to attempt to fix its failures).

And finally, the imperative of the legal-political government of the environment and climate change in the twentieth and twenty-first centuries is to control, as much as possible, the ever shifting balance between the benefit of humans (towards as wide an appreciable spread of civil peace as possible) and the maintenance of nature.

We turn now to our two promised case studies, which will further illustrate (among others) the seven crucial points.

Two brief case studies

Our first case study focuses on the Australian state of Victoria, particularly the region of Gippsland, which covers almost the entire eastern half of the state. Since 2008, litigation in the Victorian Civil and Administrative Tribunal (VCAT) relating to planning and development in Gippsland has both highlighted and generated legal uncertainty as to how strategies and regulations designed to deal with the threat of climate change will impact on the rights and obligations of councils, landowners, developers, and communities. The litigation has been an impetus to action on climate change not only because it has effectively prohibited development in coastal areas deemed most vulnerable to the effects of climate change, should it bring with it rising sea levels, but also because, in doing this, it has drawn attention to the seriousness of the threat of climate change.

Spurred on by the 2008 decision of the VCAT in the case of *Gippsland Coastal Board v South Gippsland Shire Council (No. 2)*, land-use planning has become one of the most visible ways by which the law has been used to tackle the threat of climate change in coastal areas (McDonald 2010: 14). Before this case, climate change was not a matter of great concern for Victorian planning law. In the case, VCAT was asked to review planning approval given by the South Gippsland Shire Council for six dwellings on low-lying coastal agricultural land. The Gippsland Coastal Board challenged the approval, arguing, on the basis of a 2006 scientific climate change study of the Gippsland coast, that the proposed properties would be at risk from coastal inundation caused by global warming.

The interests and views heard by the Tribunal were limited to those of the applicant (the Gippsland Coastal Board), the objector (the South Gippsland Shire Council), and any experts that each party called upon. The expert scientific prediction of significant adverse climate change, which was provided to the Tribunal by Australia's main government-sponsored science research unit, the CSIRO (Commonwealth Scientific and Industrial Research Organisation), took on a special significance in the case. It allowed VCAT to go beyond the evidence elicited in the case in order to (a) carefully consider whether it should accept the CSIRO's scientific reasoning, and (b) link this reasoning to a wider climate change discourse. While it did not formally adopt the CSIRO's findings – because they had not been subjected to any 'rigorous examination in the proceedings' – it nevertheless decided to take on board the

> general consensus that some level of climate change will result in extreme weather conditions beyond the historical record that planners and others rely on in assessing future potential impacts. It is, in our view, no longer sufficient to rely only on what has gone before to assess what may happen again in the context of coastal processes, sea levels or for that matter inundation from coastal or inland storm events.
>
> (para. 40)

In this way, VCAT thereby adopted what it saw as 'the general consensus', reasoning that the adverse effects of climate change were so 'significant' (in terms of section 60(1)(e) of the *Planning and Environment Act 1987*) as to impose an obligation on itself, for all of its land-use hearings, to consider the predicted 'environmental effect' of climate change on the proposed 'use or development' of coastal land. Ultimately, then, from VCAT's perspective, the uncertainty associated with the potential impact of climate change was of less significance than the confident consensus that inevitably there will be adverse impacts in coastal areas.

VCAT acknowledged that the obligation it has imposed upon itself opens up the possibility that it could be forced to act outside the established parameters of planning law and policy. It justified the move by reference to the fact that, 'The relevance of climate change to the planning decision making process is still in an evolutionary phase. Each case concerning possible impacts of climate change will turn on its own facts and circumstances' (para. 47).

All this is to say that the 'facts and circumstances' of the Gippsland Coastal Board case led the Tribunal to decide it was necessary to adopt the precautionary principle in place of what it saw as the business-as-usual-principle in the Shire's existing planning approval process. And because there had been no formal state policy or regulation on climate change to guide the Tribunal's deliberations, its decision took on greater significance than it otherwise might have. It led directly to the introduction of a new, broader policy framework across Victoria (de Wit and Webb 2010), mandating consideration of climate change impact in coastal areas. In doing this, the new coastal policy regime made the precautionary principle part of its policy, thereby placing significant limits on small-scale development in established coastal settlements as well as larger subdivisions on the coast outside existing settlements (Allens Arthur Robinson 2010).

In a subsequent case, *Myers v South Gippsland Shire Council (No. 2)* (2009), VCAT issued an even firmer statement of its position:

> Combined, the breadth of documents addressing climate change that are now available as background information or policies, identify that one thing is certain, the issue of climate change and the impact on coastal communities is an issue that can no longer be ignored. As decision makers we can no longer leave the issue for the next generation to sort out ... Common sense tells us that, following this approach, the Tribunal should not approve coastal developments that are likely to be unduly threatened by future flooding and/or coastal inundation, creating a mess to be dealt with by future generations.
>
> (para. 9)

From this one case study we can easily see how the precautionary principle has quickly won for itself legal and political authority, in large part because it 'does not direct a particular outcome to occur ... it regulates reasons for a decision and the process by which a decision is made' (E. Fisher 2007: 41). As de Sadeleer observes, the precautionary principle 'transforms doubt into possible certainty and hence strengthens action by public authorities in the face of uncertainty' (2002: 222). In Ewald's terms, 'Precaution finds its condition of possibility in a sort of hiatus and time-shift between the requirements of action and the certainty of knowledge' (2002: 294). The risk of legal liability, coupled with VCAT's insistence of the common sense of a precautionary approach, effectively renders otiose the uncertainty attaching to the predicted adverse effects of climate change; at least in this instance, the possible has become, for legal and policy purposes, the probable.

The mix of elements at work in this case study has law to the fore. Law here is trying, with some success, to lead politics, economic interests, and other interests. Such a mix is not unusual for legal-political government. This system of government has to be able to deal with many different mixes of its components, though its default mix is one in which politics and sovereignty try to lead law and other elements. This brings us nicely to our second case study.

It is a case study only of one fraction of an ongoing debate about land-use in a different Australian state, Queensland. This fraction of the debate has taken place across two editions of the national newspaper *The Australian*. On Saturday 1 June 2013 the paper dedicated an entire broadsheet page to two pieces about a dispute between the Queensland state government (at the time in conservative hands) and the Australian federal government (at the time in social democratic Labor Party hands). In one of these pieces, the paper's environment editor, Graham Lloyd reported on the federal government's attempts, led by Environment Minister Tony Burke, to wrest control of national parks from state governments. The issue had been brought to a head 'by emergency laws passed in Queensland to give temporary access to five national parks to graze starving cattle. Images of malnourished beasts have muddied the waters. But the weakening of park protections is of national concern.' Lloyd acknowledges that the 'federal government has no real powers to control what happens in national parks' and describes 'a fierce campaign' by 'environment groups to deal the federal government into the decision-making process ... by listing national parks as a trigger under the federal government's primary environment laws, the Environment Protection and Biodiversity Conservation Act'. He goes on to describe the alignment of interests between Minister Burke and environmental groups like the Australian Conservation Foundation and Queensland National Parks Association (QNPA) (Lloyd 2013: 17).

Where Burke is reported as stating a 'clear view that national parks are for families and nature' and that they 'are not farms, rifle ranges, mine sites or logging coupes' (the last three 'nots' are references to proposals, some of them by other conservative-led states, to allow some shooting and/or mining and/or logging in national parks), the executive co-ordinator of the QNPA is reported as insisting on a return to the management of national parks 'by the "cardinal principle", which means they will be managed for conservation first and other uses second' (Lloyd 2013: 17).

Lloyd says little about the Queensland government's position – other than to note a claim by the state's Development Minister about the need to save cattle starving to death in drought-affected areas by letting them graze in small parts of national parks and to note the government's insistence that 'the state's premier national parks are not threatened' (Lloyd 2013: 17). But the other story on the page, by Greg Roberts, is dedicated to the political machinations behind what the sub-heading to his story calls 'the greatest rollback of environmental protection in Australian political history'. The details of these machinations – to do with the complexities of relations between the two Queensland parties which recently merged into the one Liberal National Party conservative force (the Queensland Liberal Party and the once-dominant and still-powerful Queensland National Party) need not concern us here. What is important for our purposes about Roberts's story is the fact that he describes more than just the proposed land-use changes for national parks. Also on the government's agenda, he argues, are 'the scrapping of wild river declarations on Cape York' and the opening of the Cape York area 'for mining and agricultural expansion'. Like Lloyd, Roberts concentrates on presenting a range of environmentalists' objections to the governments moves (Roberts 2013: 17).

This is of course a normal and, we think, reasonable thing for a national newspaper to do – to allow its writers to develop thought-provoking arguments using careful reporting. No less normal, or less reasonable, is the paper's decision to publish a formal reply to their arguments in the following Saturday's edition, 8 June 2013. In an indication of how seriously the contributions of Lloyd and Roberts to the national debate about the environment were being treated in political circles, the reply was written by the leader of the Queensland government, Premier Campbell Newman. Newman argues that the previous Saturday's pieces, particularly that of Roberts, were too reliant on 'the environmental movement's skewed assessment of my government's genuine efforts to restore balance to Queensland's natural resource management laws and create jobs and economic opportunities in regional Queensland'. The Premier spends some time insisting that the environment is, with his government in charge, in safe hands – 'We have maintained a ban on the clearing of regrowth on the 60 per cent of the state that is leasehold land and, on freehold land, farmers will need to abide by codes of conduct when undertaking routine activities … Practices that show no regard for the environment will be detected through satellite monitoring and penalties will apply' – but it is fair to say that he spends much more time on promoting 'economic growth and allowing farmers to get on with sustainably producing food and fibre for us all' (Newman 2013: 16).

Our second case study, as we suggested it would, promotes mainly the combination of the sovereign political interests of the state and the economic interests of private landholders (and potentially private firms), though it does not discount those of law, morality, aesthetics and science. As we noted above, this mix too is not unusual for legal-political government. But what this case study highlights, especially in the contrast it provides with the first, is a point we have made a number of times, about the way the legal-political system of government encourages any number of interests to participate in the politics of any issue, in this instance the protection of the environment. We will make this matter the focus of our closing remarks.

Final thoughts

Our two case studies are good examples of that feature of the legal-political government of the environment and climate change whereby the system's commitment to the maintenance of peace, security, well-being, and prosperity delivers to the large majority of the population of the law-and-politics country involved, Australia, the opportunity to appreciate and exercise individual freedoms. These freedoms include the freedom to participate in the wide-ranging debates about different aspects of the environment and climate change. In this way, the wide-ranging debates are a manifestation of the legal-political system's 'trial and error', 'open for discussion' mode of governing. As we have said a number of times, such debates draw in many different interests expressing many different political positions.

In the two case studies presented above, the interests include those of landholders, environmentalists, regulators, political parties, newspapers, other media, small and large firms, non-government organizations, farmers, miners, shooters, and so on and so on. Their positions include stopping residential development in coastal areas, encouraging more such development, fostering economic growth in particular regions as part of a commitment to increasing prosperity, limiting economic growth in particular areas as part of commitment to increasing well-being, being cautious about certain 'sensitive' areas as part of a commitment to increasing national security, and so on and so on.

The fact that these lists could go on to become much, much longer provides an important contrast between the situation in law-and-politics countries and the situation in those countries that use governmental systems which do not encourage individual freedoms or the wide-ranging debates that go with them. In such countries, the number of interests involved in debates about the environment and climate change and the number of positions generated within the debates are bound to be much smaller than they are in any law-and-politics country. In any country which does not employ the legal-political system of government, the number of interests and the number of positions will almost certainly be restricted by the forces that make up the ruling elite in that country, which, by its nature, will seek to limit the possibility of interests outside the ruling elite becoming involved and seek to limit the competing positions in their country's debates to only a few 'acceptable' possibilities.

These points not only make it easier to make sense of the two case studies; they also make it easier to make sense of our constant claim that the legal-political government of the environment and climate change is never easy, never straightforward, and never complete or perfect. More than this, they should make it easier for the reader to recognize how we feel about the legal-political government of the environment and climate change, something we briefly touched on near the start of the book: we think the way the legal-political system of government encourages wide-ranging debate about any number of issues relating to the environment and climate change, featuring any number of interests and any number of political positions, is a very good thing. We are, this is to say, fans of the legal-political system of government (albeit fans concerned about some of the more egregious failures of the system). Whether we are fans despite this system's deliberate 'trial and error', 'never smooth sailing', 'open for discussion' nature or because of it is perhaps a matter best left for a different book.

Bibliography

ABC Online (2007a) 'Forum Hears of "High" Esperance Lead Levels', 27 March. Available: www. abc.net.au/news/newsitems/200703/s1882285.htm (accessed 6/6/2007).

ABC Online (2007b) 'Lead Contamination Hearing Extended', 12 June. Available: www. abc.net.au/news/newsitems/200706/s1948542.htm (accessed 6/6/2007).

Allens Arthur Robinson (2010) 'Environment and Planning Focus: Victorian coastal Climate Change – Issues and Options Paper 16 March 2010'. Available: www.aar. com.au/pubs/env/foenvmar10.htm (accessed 30/5/2011).

Armstrong, B. (2006) 'Breast Cancer at the ABC Toowong Queensland: Third Progress Report for the Independent Review and Scientific Investigation Panel – 21 December 2006'. Available: http://about.abc.net.au/wp-content/ uploads/2013/04/ThirdProgressReportIndependentPanelDec2006.pdf (accessed 10/6/2013).

Australian Law Reform Commission (1985) *Standing in Public Interest Litigation.* Canberra: Australian Government Publishing Service.

Barry, A., T. Osborne, and N. Rose (eds) (1996) *Foucault and Political Reason: Liberalism, Neo-Liberalism and Rationalities of Government.* London: UCL Press.

Bate, J. (1991) *Romantic Ecology: Wordsworth and the Environmental Tradition.* London: Routledge.

Beck, U. (1992) *Risk Society: Towards a New Modernity.* London: Sage.

Beck, U. (1999) *World Risk Society.* Cambridge: Polity.

Becker, C.L. (1932) *The Heavenly City of the Eighteenth-Century Philosophers.* New Haven, NJ: Yale University Press.

Bell, S. and D. McGillivray (2008) *Environmental Law*, 7th edn. Oxford: Oxford University Press.

Besant, C. (1991) 'From forest to field: A brief history of environmental law' *Legal Service Bulletin* 16: 160–64.

Billauer B., A. Moskowitz, and K. Gallinari (1989) 'The Toxic Tort is Ill: Deficiencies in the Plaintiff's Case and How to Prove Them' in D. Schnare and M. Katzman (eds) *Chemical Contamination and Its Victims: Medical Remedies, Legal Redress and Public Policy.* New York: Quorum Books, 65–84.

Bonyhady, T. (2000) *The Colonial Earth.* Melbourne: Melbourne University Press.

Bonyhady, T. (2007) 'The New Australian Climate Law' in T. Bonyhady and P. Christoff (eds) *Climate Law in Australia.* Sydney: The Federation Press, 8–31.

Bonyhady, T. and P. Christoff (eds) (2007) *Climate Law in Australia.* Sydney: The Federation Press.

Bonyhady, T., A. Macintosh, and J. McDonald (eds) (2010) *Adaptation to Climate Change: Law and Policy*. Sydney: The Federation Press.

Boughey, A. (1971) *Fundamental Ecology*. Scranton, NJ: In Text Educational Publishers.

Brewer, J. (1989) *The Sinews of Power: War, Money and the English State, 1688–1783*. London: Unwin-Hyman.

Brunnée, J., M. Doelle, and L. Rajamani (eds) (2012) *Promoting Compliance in an Evolving Climate Regime*. Cambridge: Cambridge University Press.

Burchell, D. (2002) 'Ancient Citizenship and Its Inheritors' in F.E. Isin and B. Turner (eds) *Handbook of Citizenship*. London: Sage, 89–103.

Burchell, D. (2003) 'Paradoxes of the Public Sphere: Enlightenment Fables and Digital Divides' *Southern Review* 36: 11–21.

Burchell, G., C. Gordon, and P. Miller (eds) (1991) *The Foucault Effect: Studies in Governmentality*. Brighton: Harvester Wheatsheaf.

Burns, W.C.G. and H.M. Osofsky (eds) (2009) *Adjudicating Climate Change: State, National and International Approaches*. New York: Cambridge University Press.

Calhoun, C. (ed.) (1992) *Habermas and the Public Sphere*. Cambridge, MA: MIT Press.

Campbell, D., M. Klaes, and C. Bignell (2010) 'After Cancun: The Impossibility of Carbon Trading' *University of Queensland Law Journal* 29: 163–90.

Cane, P. (1997) *The Anatomy of Tort*. Oxford: Hart.

Canning, J.P. (1988) 'Introduction: Politics, Institutions and Ideas' in J.H. Burns (ed.) *The Cambridge History of Medieval Political Thought c. 350–c. 1450*. Cambridge: Cambridge University Press, 341–66.

Cannon, M. (1998) *That Disreputable Firm … the Inside Story of Slater & Gordon*. Melbourne: Melbourne University Press.

Capriotti, S. (2002) 'Is there a Future for Cell Phone Litigation?' *Journal of Contemporary Health Law and Policy* 18: 489–510.

Carlson, A. (1998) 'Aesthetic Appreciation of the Natural Environment' in R. Botzler and S. Armstrong (eds) *Environmental Ethics: Divergence and Convergence*, 2nd edn. Boston: McGraw-Hill, 122–31.

Christie, E. (1992) 'Toxic Tort Disputes: Proof of Causation and the Courts' *Environmental and Planning Law Journal* 9: 302–18.

Churchill, W. (1947) Speech to the House of Commons *Hansard*, 11 November.

Cittadino, E. (2003) 'Ecology' in J.L. Heilbron (ed.) *The Oxford Companion to the History of Modern Science*. Oxford: Oxford University Press, 229–32.

Cole, D.H. (2011) 'From Global to Polycentric Climate Governance' *Climate Law* 2: 395–413.

Commonwealth of Australia (1992a) *Inter-governmental Agreement on the Environment*. Available: www.environment.gov.au/esd/national/igae/ (accessed 19/6/2013).

Commonwealth of Australia (1992b) *National Strategy for Ecologically Sustainable Development*. Canberra: Australian Government Publishing Service.

Conaghan, J. and W. Mansell (1993) *The Wrongs of Tort*. London: Pluto.

Coyle, S. and K. Morrow (2004) *The Philosophical Foundations of Environmental Law: Property, Rights and Nature*. Oxford: Hart.

Cranor, C. and D. Eastmond (2001) 'Scientific Ignorance and Reliable Patterns of Evidence in Toxic Tort Causation: Is there a Need for Liability Reform?' *Law and Contemporary Problems* 64: 5–48.

Dean, M. (1999) *Governmentality*. London: Sage.

Dean, M. and B. Hindess (eds) (1998) *Governing Australia: Studies in Contemporary Rationalities of Government*. Melbourne: Cambridge University Press.

De Sadeleer, N. (2002) *Environmental Principles: From Political Slogans to Legal Rules*. Oxford: Oxford University Press.

Dettelbach, M. (1996) 'Humboldtian Science' in N. Jardine, J.A. Secord, and E.C. Spary (eds) *Cultures of Natural History*. Cambridge: Cambridge University Press, 287–304.

De Wit, E. and R. Webb (2010) 'Planning for Coastal Climate Change in Victoria' *Environmental and Planning Law Journal* 27: 23–35.

Douglas, M. (1992) *Risk and Blame: Essays in Cultural Theory*. London: Routledge.

Douglas, M. and A. Wildavsky (1982) *Risk and Culture*. Oxford: Blackwell.

Du Gay, P. (2012) 'Leviathan Calling: Some Notes on Sociological Anti-statism and Its Consequences' in G. Wickham (ed.) *Sociology's Object(s) and the Discipline's Relevance*, special issue of *Journal of Sociology* 48: 397–409.

Edmond, G. and D. Mercer (2002) 'Rebels Without a Cause?: Judges, Medical and Scientific Evidence and the Uses of Causation' in I. Freckelton and D. Mendelson (eds) *Causation and Law in Medicine*. Aldershot: Ashgate, 83–124.

Edmond, G. and D. Mercer (2004) 'Daubert and the Exclusionary Ethos: The Convergence of Corporate and Judicial Attitudes towards the Admissibility of Expert Evidence in Tort Litigation' *Law and Policy* 26: 231–356.

Ewald, F. (1986) 'Social Law' in G. Teubner (ed.) *Dilemmas of Law in the Welfare State*. Berlin: Walter de Gruyter, 40–75.

Ewald, F. (1991) 'Insurance and Risk' in G. Burchell, C. Gordon, and P. Miller (eds) *The Foucault Effect: Studies in Governmentality*. London: Harvester/Wheatsheaf, 197–210.

Ewald, F. (2002) 'The Rules of Descartes's Malicious Demon: An Outline of a Philosophy of Precaution' in T. Baker and T. Simon (eds) *Embracing Risk: The Changing Culture of Insurance and Responsibility*. Chicago: University of Chicago Press, 273–301.

Farrier, D. (2007) 'The Limits of Judicial Review: Anvil Hill in the Land and Environment Court' in T. Bonyhady and P. Christoff (eds) *Climate Law in Australia*. Sydney: The Federation Press, 189–213.

Fisher, D.E. (2003) *Australian Environmental Law*. Sydney: Law Book Co.

Fisher, D.E. (2007) 'The Statutory Relevance of Greenhouse Gas Emissions in Environmental Regulation' *Environmental and Planning Law Journal* 24: 210–36.

Fisher, D.E. (2010) *Australian Environmental Law: Norms, Principles and Rules*, 2nd edn. Sydney: Law Book Co.

Fisher, E. (2007) *Risk Regulation and Administrative Constitutionalism*. Oxford: Hart.

Fleming, J. (1992) *The Law of Torts*, 9th edn. Sydney: Law Book Co.

Foucault, M. (1979) 'On Governmentality' *Ideology & Consciousness* 6: 5–21.

Freckelton, I. (2000) 'Editorial: Epidemiology Evidence' *Journal of Law and Medicine* 8: 133–41.

Furedi, F. (2009) 'Precaution Culture and the Rise of Possibilistic Risk' *Erasmus Law Review* 2: 197–220.

Gewirtz, P. (1996) 'On "I Know It When I See It"' *Yale Law Journal* 105: 1023–47.

Giddens, A. (2009) *The Politics of Climate Change*. Cambridge: Cambridge University Press.

Godden, L. and J. Peel (2010) *Environmental Law: Scientific, Policy and Regulatory Dimensions*. Melbourne: Oxford University Press.

Goodie, J. (2008) 'Toxic Tort and the Articulation of Environmental Risk' *Law Text and Culture* 12: 69–102.

Goodie, J. and G. Wickham (2002) 'Calculating "Public Interest": Common Law and the Legal Governance of the Environment' *Social and Legal Studies* 11: 37–60.

Habermas, J. (1989) [1962] *The Structural Transformation of the Public Sphere: An Inquiry Into a Category of Bourgeois Society*, trans. T. Burger, with the assistance of F. Lawrence. Cambridge, MA: MIT Press.

Hajer, M. (1995) *The Politics of Environmental Discourse: Ecological Modernization and the Policy Process*. Oxford: Clarendon.

Hanekamp, J.C., J. Vera-Navas, and S.W. Verstegen (2005) 'The Historical Roots of Precautionary Thinking: The Cultural Ecological Critique and "The Limits to Growth"' *Journal of Risk Research* 8: 295–310.

Hankinson, R.J. (1995) *The Sceptics*. London: Routledge.

Hargrove, E. (1989) *Foundations of Environmental Ethics*. Englewood Cliffs, NJ: Prentice Hall.

Harremoes, P., D. Gee, M. McGarvin, A. Stirling, J. Keys, B. Wynne, and S. Guedes Vaz (eds) (2002) *The Precautionary Principle in the 20th Century: Late Lessons from Early Warnings*. London: Earthscan.

Havemann, P. (2003) 'Genetic Modification, Ecological Good Governance and the Law: New Zealand in the Age of Risk' *James Cook University Law Review* 10: 7–50.

Helm, D. and C. Hepburn (eds) (2009) *The Economics & Politics of Climate Change*. Oxford: Oxford University Press.

Hindess, B. (1986) 'Interests in Political Analysis' in J. Law (ed.) *Power, Action and Belief*. London: Routledge & Kegan Paul, 112–31.

Hirschman, A.O. (1977) *The Passions and the Interests: Political Arguments for Capitalism before its Triumph*. Princeton, NJ: Princeton University Press.

Hobbes, T. (1845a) *The English Works of Thomas Hobbes of Malmesbury: Now First Collected and Edited by Sir William Molesworth, BART*, Volume II: *Philosophical Rudiments Concerning Government and Society {De Cive}*. London: John Bohn.

Hobbes, T. (1845b) *The English Works of Thomas Hobbes of Malmesbury: Now First Collected and Edited by Sir William Molesworth, BART*, Volume III: *Leviathan: or the Matter, Form, and Power of a Commonwealth, Ecclesiastical and Civil*. London: John Bohn.

Hobbes, T. (1845c) *The English Works of Thomas Hobbes of Malmesbury: Now First Collected and Edited by Sir William Molesworth, BART*, Volume IV: *De Corpore Politico, or the Elements of Law*. London: John Bohn.

Holder, J. and M. Lee (2007) *Environmental Protection, Law and Policy: Text and Materials*, 2nd edn. Cambridge: Cambridge University Press.

Holmes, S. (1988) 'Jean Bodin: The Paradox of Sovereignty and the Privatization of Religion' in J.R. Pennock and J.W. Chapman (eds) *Religion, Morality, and the Law*. New York: New York University Press (*Nomos* XXX), 5–45.

Holmes, S. (1994) 'Liberalism in a World of Ethnic Passions and Decaying States' *Social Research* 61: 599–610.

Hookway, C. (1990) *Scepticism*. London: Routledge.

Horwitz, M. (1982) 'The Doctrine of Objective Causation' in D. Kairys (ed.) *The Politics of Law: A Progressive Critique*. New York: Pantheon, 201–13.

Huber, P. (1991) *Galileo's Revenge: Junk Science in the Courtroom*. New York: Basic Books.

Hulme, M. (2009) *Why We Disagree About Climate Change: Understanding Controversy, Inaction and Opportunity*. Cambridge: Cambridge University Press.

Hunter, I. (1998) 'Uncivil Society: Liberal Government and the Deconfessionalisation of Politics' in M. Dean and B. Hindess (eds) *Governing Australia: Studies in Contemporary Rationalities of Government*. Cambridge: Cambridge University Press, 242–64.

Hunter, I. (2001) *Rival Enlightenments: Civil and Metaphysical Philosophy in Early Modern Germany*. Cambridge: Cambridge University Press.

Hunter, I. (2003) 'The Love of a Sage or the Command of a Superior: The Natural Law Doctrines of Leibniz and Pufendorf' in T.J. Hochstrasser and P. Schröder (eds) *Early Modern Natural Law Theories: Contexts and Strategies*. Berlin: Kluwer, 169–94.

Hunter, I. (2005) 'Security: The Default Setting of the Liberal State' *Australian Policy Online*, 7 November. Available: www.apo.org.au/webboard/results. chtml?filename_num=42404 (accessed 25/3/2013).

Hunter, I. (2006) 'The History of Theory' *Critical Inquiry* 33: 78–112.

Hunter, I. (2009) 'Spirituality and Philosophy in Post-structuralist Theory' *History of European Ideas* 35: 265–75.

Hunter, I. (2010) 'Natural Law as Political Philosophy' in D. Clarke and C. Wilson (eds) *The Oxford Handbook of Philosophy in Early Modern Europe*. Oxford: Oxford University Press, 475–99.

Hunter, I. (2012) 'The Figure of Man and the Territorialisation of Justice in "Enlightenment" Natural Law: Pufendorf and Vattel' in A. Cook and N. Curthoys (eds) *Discourses of Humanity in the Enlightenment*, special issue of *Intellectual History Review* 23: 1–19.

Hunter, I., T. Ahnert, and F. Grunert (2007) 'Introduction' in C. Thomasius *Christian Thomasius: Essays on Church, State, and Politics*, ed., trans., intro. I. Hunter, T. Ahnert, and F. Grunert. Indianapolis, IN: Liberty Fund, ix–xxiii.

Hunter, I. and D. Saunders (2003a) 'Bringing the State to England: Andrew Tooke's Translation of Samuel Pufendorf's *De officio hominis et civis*' *History of Political Thought* 24: 218–34.

Hunter, I. and D. Saunders (2003b) 'Introduction' in S. Pufendorf *Samuel Pufendorf: The Whole Duty of Man, According to the Law of Nature. Together with two discourses and a commentary by Jean Barbeyrac*, ed., trans, and notes I. Hunter and D. Saunders. Indianapolis, IN: Liberty Fund, ix–xvii

Hutton, D. and L. Connors (1999) *A History of the Australian Environment Movement*. Cambridge: Cambridge University Press.

Jasanoff, S. (1995) *Science at the Bar: Law, Science and Technology in America*. Cambridge, MA: Harvard University Press.

Jasanoff, S. (2010) 'A New Climate for Society' *Theory, Culture & Society* 27: 233–53.

Kant, I. (1970) *Kant: Political Writings*, ed. H. Reiss, trans. H.B. Nisbet. Cambridge: Cambridge University Press.

Kant, I. (1996) *Practical Philosophy*, ed., trans. M.J. Gregor. Cambridge: Cambridge University Press.

Kellow, A. (1999) *International Toxic Risk Management: Ideals, Interests and Implementation*. Cambridge: Cambridge University Press.

Kerr-Forsyth, H. (2005) 'Macarthurs' Grand Park' *The Weekend Australian*, September 3–4: 64.

Koselleck, R. (1988) *Critique and Crisis: Enlightenment and the Pathogenesis of Modern Society*. Cambridge, MA: MIT Press.

Koskenniemi, M. (2001) *The Gentle Civilizer of Nations: The Rise and Fall of International Law 1870–1960*. Cambridge: Cambridge University Press.

Kriegel, B. (1995) *The State and the Rule of Law*, trans. M.A. LePain and J.C. Cohen. Princeton, NJ: Princeton University Press.

Kroll-Smith, S. and S. Westervelt (2004) 'People, Bodies and Biospheres: Nexus and the Toxic Tort' *Law and Policy* 26: 177–86.

Kubasek, N. and G. Silverman (2013) *Environmental Law*, 8th edn. Upper Saddle River, NJ: Pearson Prentice Hall.

Lanthier, I. and L. Olivier (1999) 'The Construction of Environmental "Awareness"' (trans. M. Eloy) in E. Darier (ed.) *Discourses of the Environment*. Oxford: Blackwell, 63–78.

Laudan, R. (2003) 'Observation and Experiment' in J.L. Heilbron (ed.) *The Oxford Companion to the History of Modern Science*. Oxford: Oxford University Press, 593–94.

Lee, M. (2005) *EU Environmental Law*. Oxford: Hart Publishing.

Lee, R. (2000) 'From the Individual to the Environmental: Tort Law in Turbulence' in J. Lowry and R. Edmunds (eds) *Environmental Protection and the Common Law*. Oxford: Hart, 77–91.

Lin, J. (2012) 'Climate Change and the Courts' *Legal Studies* 32: 35–57.

Lloyd, G. (2013) 'Clock Ticking for Tony Burke to Save National Parks' *The Australian* 1 June, 17.

Lord, R., S. Goldberg, L. Rajamani, and J. Brunnée (eds) (2012) *Climate Change Liability: Transnational Law and Practice*. Cambridge: Cambridge University Press.

Loughlin, M. (2003) *The Idea of Public Law*. Oxford: Oxford University Press.

Loughlin, M. (2009) 'In Defence of *Staatslehre*' *Der Staat* 48: 1–27.

Lupton, D. (1999) *Risk*. London: Routledge.

Lyster, R. (2007) 'Chasing Down the Climate Change Footprint of the Public and Private Sectors: Forces Converge Part II' *Environmental and Planning Law Journal* 24: 450–79.

Machiavelli, N. (1979) 'The Prince' in P. Bondanella and M. Musa (eds) *The Portable Machiavelli*, trans. P. Bondanella and M. Musa. New York: Penguin, 77–166.

McBarnet, D. (1981) *Conviction: Law, the State, and the Construction of Justice*. London: Macmillan.

McCalman, I. (2009) *Darwin's Armada: Four Voyages and the Battle for Evolution*. London: W.W. Norton.

McDonald, J. (2010) 'Mapping the Legal Landscape of Climate Change Adaptation' in T. Bonyhady, A. Macintosh, and J. McDonald (eds) *Adaptation to Climate Change: Law and Policy*. Sydney: The Federation Press, 1–37.

McGrath, C. (2007) 'The Xstrata case: Pyrrhic Victory or Harbinger?' in T. Bonyhady and P. Christoff (eds) *Climate Law in Australia*. Sydney: The Federation Press, 214–29.

McHugh, M. (1989) 'Neighbourhood Proximity and Reliance' in P. Finn (ed.) *Essays on Damages*. Sydney: Law Book Co., 5–42.

McMichael, A. (2001) *Human Frontiers, Environments and Disease: Past Patterns, Uncertain Futures*. Cambridge: Cambridge University Press.

Melton, J. Van Horn (2003) *The Rise of the Public in Enlightenment Europe*, Cambridge: Cambridge University Press.

Mendelsohn, A. (2003) 'Bacteriology and Microbiology' in J.L. Heilbron (ed.) *The Oxford Companion to the History of Modern Science*. Oxford: Oxford University Press, 75–77.

Miller, P. and N. Rose (1990) 'Governing Economic Life' *Economy and Society* 19: 1–31.

Morrow, K. (2000) 'Nuisance and Environmental Protection' in J. Lowry and R. Edmunds (eds) *Environmental Protection and the Common Law*. Oxford: Hart, 139–59.

Muir, J. (1998) 'A Near View of the High Sierra' in R. Botzler and S. Armstrong (eds) *Environmental Ethics: Divergence and Convergence*, 2nd edn. Boston: McGraw Hill, 108–14.

Mullany, N. (1997) 'Fear for the Future: Liability for Infliction of Psychiatric Damage' in N. Mullany (ed.) *Torts in the Nineties*. Sydney: LBC Information Services, 101–73.

Newman, C. (2013) 'Let's Get Facts on LNP and Rural Land' *The Australian* 8 June, 16.

O'Malley, P. (1999a) 'Consuming Risks: Harm Minimisation and the Government of "Drug Users"' in R. Smandych (ed.) *Governable Places: Readings in Governmentality and Crime Control*. Aldershot: Dartmouth, 191–214.

O'Malley, P. (1999b) 'Imagining Insurance: Risk, Thrift and Industrial Life Insurance in Britain' *Connecticut Insurance Law Journal* 5: 676–705.

O'Malley, P. (2000) 'Uncertain Subjects: Risks, Liberalism and Contract' *Economy and Society* 29: 459–84.

O'Malley, P. (2004) *Risk, Uncertainty and Government*. London: The Glasshouse Press.

Osborne, T. (1998) *Aspects of Enlightenment: Social Theory and the Ethics of Truth*. London: UCL Press.

Osler, M.J. (ed.) (1991) *Atoms, Pneuma, and Tranquility: Epicurean and Stoic Themes in European Thought*. Cambridge: Cambridge University Press.

Oxford English Dictionary, online edition. Available: www.oed.com (entry 63089) (accessed 23/3/2013).

Oxford English Dictionary, online edition. Available: www.oed.com (entry 59380) (accessed 4/4/2013).

Pancaldi, G. (2003) 'Biology' in J.L. Heilbron (ed.) *The Oxford Companion to the History of Modern Science*. Oxford: Oxford University Press, 92–95.

Parker, J. (2013) *Global Crisis: War, Climate Change and Catastrophe in the Seventeenth Century*. New Haven, CT: Yale University Press.

Peel, J. (2005) *The Precautionary Principle in Practice: Environmental Decision-making and Scientific Uncertainty*. Sydney: The Federation Press.

Peel, J. (2007) 'The Role of Climate Litigation in Australia's Response to Global Warming' *Environmental and Planning Law Journal* 24: 90–105.

Peel, J. (2008) 'Climate Change Law: The Emergence of a New Legal Discipline' *Melbourne University Law Review* 32: 922–79.

Peel, J. and H.M. Osofsky (2013) 'Climate Change Litigation's Regulatory Pathways: A Comparative Analysis of the United States and Australia' *Law and Policy* 35: 150–83.

Popkin, R.H. (1968) *The History of Scepticism from Erasmus to Descartes*. New York: Harper Torchbooks.

Popper, K. (1959) *The Logic of Scientific Discovery*. New York: Basic Books.

Prins, G., I. Galiana, C. Green, R. Grundmann, M. Hulme, A. Korhola, F. Laird, T. Nordhaus, R. Pielke Jnr, S. Rayner, D. Sarewitz, M. Shellenberger, N. Stehr, and H. Tezuka (2010) *The Hartwell Paper: A New Direction for Climate Policy after the Crash of 2009*. Institute for Science Innovation and Society, University of Oxford and LSE Mackinder Programme, May 2010. Available: http://eprints.lse. ac.uk/27939/1/HartwellPaper_English_version.pdf (accessed 10/6/2013).

Prins, G. and S. Rayner (2007) 'Commentary: Time to Ditch Kyoto' *Nature* 449: 973–75.

Pyenson, L. and Sheets-Pyenson, S. (1999) *Servants of Nature: A History of Scientific Institutions, Enterprises, and Sensibilities*. New York: W.W. Norton.

Rabin, R. (1993) 'Institutional and Historical Perspectives on Tobacco Tort Liability' in R. Rabin and S. Sugarman (eds) *Smoking Policy: Law, Politics and Culture*. New York: Harper Torchbooks, 111–30.

Rabin, R. (2001) 'The Tobacco Litigation: A Tentative Assessment' *DePaul Law Review* 51: 331–58.

Raymond, J. (2008) 'Nedham [Needham], Marchamont' *Oxford Dictionary of National Biography*, online edition. Available: www.oxforddnb.com/index/19/101019847 (accessed 25/3/2013).

Rice, S. (1993) 'Redefining the Public Interest' *Alternative Law Journal* 18: 191–93.

Roberts, G. (2013) 'Campbell Newman's LNP bulldozing pre-election promise' *The Australian* 1 June, 17.

Rose, N. (1989) *Governing the Soul: The Shaping of the Private Self*. London: Routledge.

Rose, N. (1993) 'Government, Authority and Expertise in Advanced Liberalism' *Economy and Society* 22: 283–99.

Rose, N. (1996) 'The Death of the Social? Refiguring the Territory of Government' *Economy and Society* 25: 327–56.

Rose, N. and P. Miller (1992) 'Political Power Beyond the State: Problematics of Government' *British Journal of Sociology* 43: 173–205.

Rudd, K. (2007) 'Climate Change: The Great Moral Challenge of Our Generation', address delivered to the National Climate Change Summit, Saturday 31 March. Available: http://trove.nla.gov.au/work/38337288?q&versionId=50805471 (accessed 20/6/2013).

Rutherford, P. (1999) 'Ecological Modernization and Environmental Risk' in E. Darier (ed.) *Discourses of the Environment*. Oxford: Blackwell, 95–118.

Rutherford, P. (2000) 'Environmental Education', a contribution to the 'History of the Present' email list: isin@calumet.yorku.ca, 1 October.

Sands, P., J. Peel, A. Fabra, and R. McKenzie (2012) *Principles of International Environmental Law*, 3rd edn. Cambridge: Cambridge University Press.

Saul, B., S. Sherwood, J. McAdam, T. Stephens, and J. Slezak (2012) *Climate Change in Australia: Warming to the Global Challenge*. Sydney: Federation Press.

Saunders, D. (1997) *Anti-lawyers: Religion and the Critics of Law and State*. London: Routledge.

Saunders, D. (2002) '"Within the Orbit of This Life" – Samuel Pufendorf and the Autonomy of Law' *Cardozo Law Review* 23: 2173–98.

Sauter, M.J. (2004) 'Preaching, a Ponytail, and an Enthusiast: Rethinking the Public Sphere's Subversiveness in Eighteenth-Century Prussia' *Central European History* 37: 544–67.

Schilling, H. (1986) 'The Reformation and the Rise of the Early Modern State' in J.D. Tracey (ed.) *Luther and the Modern State in Germany*. Kirksville, MO: Sixteenth Century Journal Publishers, 21–30.

Schmitt, C. (1976) [1927] *The Concept of the Political*, trans., notes, intro. G. Schwab, afterword L. Strauss. Princeton, NJ: Princeton University Press.

Schmitt, C. (2005) [1922] *Political Theology: Four Chapters on the Concept of Sovereignty*, trans., intro. G. Schwab, foreword T.B. Strong. Chicago: University of Chicago Press.

Scott, J. (ed.) (2009) *Environmental Protection: European Law and Governance*. Oxford: Oxford University Press.

Shaw, C. (2010) 'The Dangerous Limits of Dangerous Limits: Climate Change and the Precautionary Principle' *Sociological Review* 58: 103–123.

Skinner, Q. (2002a) *Visions of Politics, Volume 2: Renaissance Virtues*. Cambridge: Cambridge University Press.

Skinner, Q. (2002b) *Visions of Politics, Volume 3: Hobbes and Civil Science*. Cambridge: Cambridge University Press.

Stallworthy, M. (2009) 'Legislating Against Climate Change: A UK Perspective on a Sisyphean Challenge' *Modern Law Review* 72: 412–36.

Stapelton, J. (1995) 'Tort, Insurance and Ideology' *Modern Law Review* 58: 820–45.

Steele, J. (2001) 'Participation and Deliberation in Environmental Law: Exploring a Problem-solving Approach' *Oxford Journal of Legal Studies* 21: 415–42.

Steele, J. (2004) *Risks and Legal Theory*. Oxford: Hart.

Swan, N. (2007) 'The Inside Story – Breast Cancer at the ABC's Brisbane Offices' *The Health Report*. Available: www.abc.net.au/rn/healthreport/stories/2007/1838331. htm (accessed 26/2/2007).

Taylor, W. (2004) *The Vital Landscape: Nature and the Built Environment in Nineteenth-Century Britain*. Aldershot: Ashgate.

The West Australian (2007) 'Compo call over port lead scandal', 5 May, 8.

Thomas, K. (1983) *Man and the Natural World: Changing Attitudes in England 1500–1800*. London: Penguin.

Thoreau, H. (1998) 'Walking' in R. Botzler and S. Armstrong (eds) *Environmental Ethics: Divergence and Convergence*, 2nd edn. Boston: McGraw-Hill, 99–108.

Toffolon-Weiss, M. and J. Timmons Roberts (2004) 'Toxic Torts, Public Interest Law and Environmental Justice: Evidence from Louisiana' *Law and Policy* 26: 259–87.

Turner, S.P. (2002) 'Weber, the Chinese Legal System, and Marsh's Critique' *Comparative & Historical Sociology* (Newsletter of the ASA Comparative and Historical Sociology section) 14: 1–4.

Turner, S.P. and R.A. Factor (1987) 'Decisionism and Politics: Weber as Constitutional Theorist' in S. Lash and S. Whimster (eds) *Max Weber, Rationality and Modernity*. London: Allen & Unwin, 334–54.

Valverde, M., R. Levi, and D. Moore (2005) 'Legal Knowledges of Risk' in Law Commission of Canada (ed.) *Law and Risk*. Vancouver, BC: UBC Press, 86–120.

Van Caenegem, R. (1988) 'Government, Law and Society' in J.H. Burns (ed.) *The Cambridge History of Medieval Political Thought c. 350–c. 1450*. Cambridge: Cambridge University Press, 174–210.

Vanhala, L. and C. Hilson (eds) (2013) *Climate Change Litigation Symposium*, special issue of *Law and Policy*. 35: 141–260.

Walter, R. (2008a) 'Governmentality Accounts of the Economy: A Liberal Bias?' *Economy & Society* 37: 94–114.

Walter, R. (2008b) 'The Economy and Pocock's Political Economy' *History of European Ideas* 34: 334–44.

Walter, R. (2011) *A Critical History of the Economy: On the Birth of National and International Economies*. London: Routledge.

Walters, W. (2012) *Governmentality: Critical Encounters*. London: Routledge.

Weir, L. (1996) 'Recent Developments in the Governance of Pregnancy' *Economy and Society* 25: 372–92.

Wickham, G. (2006) 'Foucault, Law and Power: A Reassessment' *Journal of Law and Society* 23: 596–614.

Wickham, G. (2008) 'The Social Must Be Limited: Some Problems with Foucault's Approach to Modern Positive Power' *Journal of Sociology* 44: 29–44.

Wickham, G. (2010) 'Sociology, the Public Sphere, and Modern Government: A Challenge to the Dominance of Habermas' *British Journal of Sociology* 61: 155–75.

Wickham, G. and D. Bryan (2012) 'Money, Post-Crisis Financial Regulation, and the Fragility of Civil Peace: Maintaining Order in the Face of Chaos' *Griffith Law Review* 21: 190–208.

Wordsworth, W. (1888) 'The Excursion' in *The Complete Poetical Works*. Available: www.bartleby.com/145/. (accessed 20/6/2013).

Worster, D. (1994) *Nature's Economy: The Roots of Ecology*, 2nd edn. Cambridge: Cambridge University Press.

Zahar, A., J. Peel and L. Godden (2013) *Australian Climate Law in Global Context*. Cambridge: Cambridge University Press.

Cases cited

Aldous v Greater Taree City Council and Another [2009] NSWLEC 17.

American Electric Power Co. v Connecticut 131 S. Ct 2527 (2011).

Australian Conservation Foundation Inc v Commonwealth (1980) 146 CLR 493.

Australian Conservation Foundation v Latrobe City Council (2004) 140 LGERA 100.

Australian Conservation Foundation v Minister for Resources (1989) 19 ALD 70.

E.M. Baldwin & Son Pty Ltd v Plane & Anor; Jsekarb Pty Ltd v Plane & Anor (1999) AustTortsR 81–499.

Boyce v Paddington Borough Council (1903) 1 Ch 109, 114.

Chappel v Hart (1998) 195 CLR 232.

Charles & Howard Pty Ltd v Redland Shire Council (2008) QCA 200.

Coalition for Responsible Regulation v Environment Protection Agency 684 F.3d 102 (D.C. Cir. 2012).

Daubert v Merrell Dow Pharmaceuticals, Inc. 509 US 579, 113 SCt 2786, 125 LEd2d 469 (1993).

Drake Brockman v Minister for Planning (2007) NSWLEC 490.

Gippsland Coastal Board v South Gippsland Shire Council (No. 2) [2008] VCAT 1545.

Graham and Graham v Re-Chem [1996] Env. LR 158.

Gray v Minister for Planning [2006] NSWLEC 720.

Greenpeace Australia Ltd v Redbank Power Company (1994) 84 LGERA 143.

Hanrahan v Merck, Sharp & Dohme (Ireland) Ltd [1988] ILRM 629.

Kivalina v Exxonmobil Corporation, et al 663 F Supp 2d 863 (ND Cal Sept 30, 2009).

Massachusetts v Environmental Protection Agency 549 U.S. 583 (2007).

Minister for Planning v Walker (2008) 161 LGERA 423.

Myers v South Gippsland Shire Council (No. 2) (2009) VCAT 2414.

Napolitano v CSR Ltd & Anor (1994) (Unreported, Supreme Court of Western Australia, Seaman J, 30 August 1994) Library NO 94087.

Newman v Motorola Inc. 218 F Supp 2d 769 (D Md 2002).

Newman v Motorola Inc. No. 02-2424 Court of Appeals (4th Cir 2003).

Northcape Properties v District Council of Yorke Peninsula [2008] SASC 57.

Ogle v Strickland (1987) 71 ALR 4.

Oshlack v Richmond River Council [1994] NSWLEC 20 (25 February 1994).

Oshlack v Richmond River Council [1998] HCA 11 (25 February 1998).

Queensland Conservation Council Inc v Xstrata Coal Queensland Pty Ltd [2007] QCA 338.

Re United States Tobacco Company and Australian Federation of Consumer Organisations Inc. and: The Minister for Consumer Affairs; The Trade Practices Commission and Australian Federation of Consumer Organisations (1988) 83 ALR 79.

Rylands v Fletcher (1868) LR 3 HL 330.

Seltsam Pty Ltd. v McGuiness (2000) 49 NSWLR 262.

Sturges v Bridgeman (1879) 11 Ch D 852.

Taralga Landscape Guardians v Minister for Planning [2007] NSWLEC 59.

Tipping v St Helens Smelting Company (1865) LR 1 Ch App 66.

Wildlife Preservation Society of Queensland Proserpine/Whitsunday Branch Inc v Minister for Environment [2006] FCA 736.

Index